Mechanism of Saturated Soil-Tunnel Interaction under Medium Blast Loading
爆炸荷载条件下土与隧道相互作用机理研究

韩玉珍 张 雷 华福才 著

中国建筑工业出版社

图书在版编目（CIP）数据

Mechanism of Saturated Soil-Tunnel Interaction under Medium Blast Loading=爆炸荷载条件下土与隧道相互作用机理研究：英文/韩玉珍，张雷，华福才著. —北京：中国建筑工业出版社，2021.10
ISBN 978-7-112-26731-6

Ⅰ.①M… Ⅱ.①韩…②张…③华 Ⅲ.①岩土工程-研究-英文 Ⅳ.①TU4

中国版本图书馆 CIP 数据核字（2021）第 211737 号

责任编辑：辛海丽
责任校对：赵 菲

Mechanism of Saturated Soil-Tunnel Interaction under Medium Blast Loading
爆炸荷载条件下土与隧道相互作用机理研究
韩玉珍 张 雷 华福才 著
*
中国建筑工业出版社出版、发行（北京海淀三里河路 9 号）
各地新华书店、建筑书店经销
北京科地亚盟排版公司制版
北京建筑工业印刷厂印刷
*

开本：787 毫米×1092 毫米 1/16 印张：5¾ 字数：142 千字
2021 年 10 月第一版　2021 年 10 月第一次印刷
定价：36.00 元
ISBN 978-7-112-26731-6
(38055)

版权所有　翻印必究
如有印装质量问题，可寄本社图书出版中心退换
（邮政编码 100037）

Foreword

A series of numerical simulations were carried out in this research to study on the interaction between subway tunnels and saturated soils subjected to medium internal blast loading (<200kg of TNT-equivalent). Numerical procedure using Finite Element program LS-DYNA that can model explosion, air-solid interaction, material damage and soil-structure interaction were used to conduct the simulations. The excess pore water pressure was studied with an existing soil model (FHWA) which can simulate pore water pressure and effective soil pressure. A recently developed blast loading scheme which removes the necessity to model the explosive in the numerical models but still maintains the advantages of nonlinear fluid-structure interaction was used to study the process of blast wave propagation in the air domain inside the tunnel.

Centrifuge technique, which has been successfully employed in model tests to investigate the blast effects, was used to simulate the effect of medium- to large-scale explosion at small scaled distance. Extensive parametric studies were carried out to understand the failure and damage of cast-iron subway tunnels under internal blast loading and to shed light on their protection by structural and geotechnical measures.

The study found that the blast loading technique could significantly reduce the computation effort and the initial density of air in the numerical model could be artificially increased to partially compensate the error induced by the use of relatively large air elements. And as long as the free air blast at a specific scaled distance was properly simulated, the fluid-structure interaction could be properly duplicated using proper ALE coupling scheme.

Increase of soil stiffness by certain soil reinforcement technique, increasing overburden stress on the tunnel and its surrounding soil without significantly increasing static lining stress, and enhancing lining damping through structural measures at its joints, may be adopted to reduce the vulnerability of tunnel. Protective measures by geofoam may only be effective for small amount of explosive. Explosion close to ventilation shafts would reduce the blast pressure on the lining, but might not alleviate lining damage due to the reduction of lining integrity.

Contents

Chapter 1 Introduction and Literature Review ··· 1
  1.1 Background and Research Objective ·· 1
  1.2 Literature Review ·· 3
    1.2.1 Responses of Underground Structures Subjected to Blast Loading ············ 3
    1.2.2 Pore Pressure and Soil Liquefaction Induced by Blast ···························· 4
    1.2.3 Multi-hazard Engineering for Tunnel Lining ·· 4
    1.2.4 Centrifuge Modeling Technique ·· 5
    1.2.5 Numerical Simulation of Explosion ·· 5
    1.2.6 Numerical Simulation of Soil-Tunnel Interaction ·································· 6
  1.3 Research Approaches ·· 7
  1.4 Chapter Contents ·· 7

Chapter 2 Blast Simulation Involving Air-Structure Interaction ··················· 9
  2.1 Blast Wave Propagation in Air ·· 9
    2.1.1 Theoretical Background ·· 9
    2.1.2 Blast Wavelength and Finite Element Size ··· 10
  2.2 Simulation of Blast Loading in LS-DYNA ·· 11
    2.2.1 Modeling of Air and Explosives ·· 11
    2.2.2 MM-ALE Approach ··· 13
    2.2.3 LBE+MM-ALE Approach ·· 14
  2.3 Simulation of Free Air Blast ·· 16
    2.3.1 Effects of Initial Air Density ·· 17
    2.3.2 Simulation of Free Air Blast with Different Locations of Ambient Layer ····· 17
    2.3.3 Amount of Explosive and Element Size ·· 18
  2.4 Centrifuge Technique ·· 20
  2.5 Reflected Blast on a Rigid Surface ··· 20
  2.6 Simulation of Blast Effect on a Steel Plate ······································· 21
  2.7 Size of Air Domain Inside the Tunnel ··· 23
  2.8 Conclusions and Discussions ··· 25

Chapter 3 Numerical Modeling of Soil-Tunnel Interaction ·························· 27
  3.1 Modeling of Tunnel Lining ··· 27
  3.2 Modeling of Soil Tunnel Interfaces ··· 28
  3.3 Modeling of Unsaturated and Saturated Soils ··································· 29
  3.4 Evaluation of FHWA Soil Model ·· 30

3.4.1　Introduction ································· 30
　　　3.4.2　Evaluation ································· 31
　3.5　Conclusions ································· 37
Chapter 4　Response of Circular Tunnels Subjected to Medium Internal Blast Loading
　　································· 38
　4.1　Base-Case Finite Element Model ································· 38
　4.2　Simulation Results with 200kg TNT ································· 40
　　　4.2.1　General Response of Tunnel ································· 40
　　　4.2.2　Response in Soil-Tunnel Interface ································· 43
　　　4.2.3　Response in Soil ································· 45
　　　4.2.4　Analysis Results with Simplified Blast Load ································· 48
　　　4.2.5　Equivalent Triangle Impact Load ································· 49
　4.3　Simulation Results with 100kg TNT ································· 51
　　　4.3.1　Response of Tunnel ································· 51
　　　4.3.2　Impact Loading on Tunnel Lining ································· 54
　　　4.3.3　Response in Soil ································· 55
　4.4　Simulation Results with 50kg TNT ································· 57
　　　4.4.1　Response of Tunnel ································· 57
　　　4.4.2　Response of Soil ································· 58
　4.5　Conclusions ································· 60
Chapter 5　Parametric Study ································· 62
　5.1　Lining Strength and Stiffness ································· 62
　5.2　Bulk modulus K of Saturated Soil ································· 64
　5.3　Liquefaction Susceptibility of Saturated Soil under Blast Loading ················ 65
　5.4　Burial Depth ································· 68
　5.5　Thickness of Tunnel Lining ································· 70
　5.6　Damping of Tunnel Lining ································· 71
　5.7　Influence of Ventilation Shaft ································· 72
　5.8　Influence of Foam Liner ································· 76
　5.9　Conclusion ································· 79
Chapter 6　Conclusions and Future Research ································· 81
　6.1　Conclusions ································· 81
　6.2　Future Research ································· 82
Bibliography ································· 83

Chapter 1 Introduction and Literature Review

1.1 Background and Research Objective

US transportation system has 337 highway tunnels and 211 transit tunnels in 2003 according to the Blue Ribbon Panel (BRP) on Bridge and Tunnel Security assigned by AASHTO. These tunnels are facing threats of internal explosion which will cause large socioeconomic losses. The new technology should be developed to meet the need of tunnel safety. This research aims to facilitate this target through study on the interaction between circular subway tunnels and saturated soils subjected to medium internal blast loading.

These tunnels have become one of the targets of terrorist attacks in recent years and bombing is one main scheme. The February 2004 Moscow metro bombing killed 41 people on the Zamoskvoretskaya Line, up to 120 people were injured in the incident, some of the more common injuries being broken bones and smoke inhalation. The 2010 Moscow metro bombings at two stations of the Moscow metro killed at least 40 people and over 100 injured. From July to October in 1995, a series of bombing attacks happened in Paris metro systems. In total, these attacks killed 8 people and injured more than 100. The biggest one happened on July 25, 1995, a gas bottle exploded in the Saint-Michel station of line B of the RER (Paris regional train network), killing 8 people and wounding 80.

Therefore the preventive measures should be developed to reduce the possibility of bombing attacks and avoid collapse of existing subway structures. The internal blast load should be taken into consideration in the design. However the guidelines to design underground structures considering the internal blast loading are still lacking. One reason is that the bombing attacks are totally disasters after the explosive explodes. More measures are taken to investigate the potential explosion. If traditional protective measures are applied on the whole structures system, the expense could be too high. The other reason is the complicated characteristics of the problem. This issue involves coupled fluid-solid interaction, dynamic soil-tunnel interactions, structure damage and nonlinear response of soil.

Many transportation tunnels run through saturated soils and blast loading of saturated soils induces drastic changes of compressive strain and excess pore pressure. Especially the residue of excess pore pressure will reduce effective stress and may result in soil liquefaction. The developments of compressive strain, pore pressure and effective stress interact with the transportation tunnel and internal blast pressure complicate the tunnel response and damage. Figure 1.1 shows the particle-velocity profile surrounding a 5m diameter cast-iron tunnel subjected to an internal explosion of 75kg TNT via numerical simulation. The soil has blast induced liquefaction when the soil particle velocity is larger than 0.4cm/s. However, mechanism of saturated soil-tunnel interaction under medium internal blast loading is still not well understood. There is no validated design guideline for tunnel linings in saturated soil due to medium blast loading.

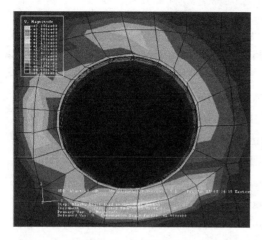

Figure 1.1 Soil particle velocity profile due to explosion of 75kg TNT inside a 5m diameter tunnel with 7cm cast-iron lining (unit: m/s; the smallest velocity shown is 0.4cm/s)

Some transportation agencies such as Port Authority of NY and NJ carried out research to identify the blast vulnerability of specific tunnels but the results are not available to the public due to security concerns. More critically, general mechanism other than specific response is necessary for the hazard analysis of underground tunnels in a multi-hazard environment, which may not be readily obtained even if their findings are released.

The existing knowledge on explosion-induced compressive strain and pore pressure in saturated soil does not directly apply herein because explosion does not occur in soil and the tunnel lining modifies the characteristics of blast loading on soil. The vibration of tunnel after explosion introduces multiple shocks instead of one blast, and the soil particle velocity from internal explosion with medium amount of explosive may still be large enough to generate significant excess pore pressure. Analyzing the effects of blast on tunnels and soils is a difficult task, as it involves highly nonlinear fluid dynamics, structural dynamics and fluid-structure interaction. At present, most blast resistant analyses make use of simplistic blast loadings and structure models, but their accuracy cannot be guaranteed when complicated structure and loading scenarios are involved. For example, it is difficult to assume a rational blast loading for the underground tunnel subjected to internal blast. The existing blast loading equations are focused on the free air blast, blast effect on plane rigid surface, or blast effect inside a rectangular structure. The effects of confinement from tunnels on the blast loading are not studied or investigated thoroughly.

The research hopes to shed lights on the following questions:

(a) Under medium internal blast, what is the damage mode of tunnels in saturated soil?

(b) What is the extent of residual excess pore pressure as well as its effect on tunnel response?

(c) What is the relation between saturated soil-tunnel interaction and soil parameters?

(d) What are the effective mitigation measures to protect the tunnels from medium internal blast loading?

These answers are valuable and will contribute to improve the design and rehabilitation of transportation tunnels in saturated soils considering multiple hazards. The subject of the study is single-track cast-iron subway tunnel in saturated soil. Dense saturated soil was considered as confining media surrounding subway structures. Single-track cast iron tunnel lining is used extensively in the subway system in New York City. The common equivalent explosive targeting the subway tunnels is 50 to 200kg TNT from the recommendations of Federal Emergency Management Agency (FEMA 2003).

1.2 Literature Review

1.2.1 Responses of Underground Structures Subjected to Blast Loading

Responses of underground structures subjected to explosive loading have been extensively studied for its military importance. For explosions outside underground structures, most of the studies focused on cratering, earth pressure on underground structures, and corresponding structure damage.

Only few of these studies considered the coupling of pore fluid and soil particles, not to mention the change of effective stress and its effect on underground structures.

For explosions inside underground structures, air-blast, ground blast wave, blast pressure, collapse and debris of underground structures have been investigated. These studies are mostly related to large-scale explosions inside underground ammunition storages in rock mass, the findings of which cannot be directly applied to tunnels in saturated soils.

Few studies on the responses of underground structures subjected to internal blast loading can be found. The subjects of Chille et al. and Choi et al. were both underground structures in rock masses. Choi et al., through analysis of coupled air-solid interaction, found that the blast pressure on tunnel lining was not the same as the CONWEP normally reflected pressure. Preece et al. investigated the response of a 13ft diameter aluminum tunnel in moist soil subjected to internal

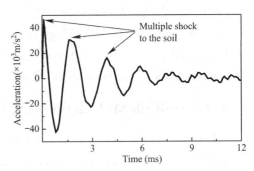

Figure 1.2 Acceleration of cast-iron lining (6.5cm thick, 5m inner diameter) subject to internal explosion of 75kg TNT

blast loading from 6600pounds of TNT using centrifuge test, which is not realistic for the hazard facing general transportation tunnels. Port Authority of New York and New Jersey and several other transportation agencies investigated the blast vulnerability of specific tunnels after "9 • 11" but unfortunately their results are not released. Very recently, through numerical analysis it is found that single track subway tunnels in saturated silty soil with cast-iron lining, which are used extensively in New York City, would damage under modest internal explosion (50~75kg of TNT-equivalent). This research also found that under single blast loading, the tunnel vibrated drastically and applied multiple shocks to the ground media, which coincided with the finding of Feldgun et al. as shown in Figure 1.2.

1.2.2 Pore Pressure and Soil Liquefaction Induced by Blast

Under blast loading, the large compressive pressure can induce large compressive strain owing to the compression of soil particles and pore water, and upon unloading, the inelastic compressibility of soil skeleton would induce residual excess pore pressure, the magnitude of which can be adequately large to liquefy saturated soils, as illustrated in Figure 1.3. This issue has been investigated extensively since 1960's and blast loading has been used to densify loose soil deposits and to initiate soil liquefaction for research purpose.

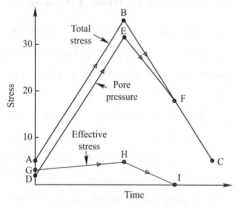

Figure 1.3 Illustration of compressional soil liquefaction (from Fragazsy and Voss)

Studies have found that more crushable soil particles result in higher compressive strain and residual excess pore pressure. While most studies have focused on sandy soils, some showed that silty soil and clayey sand can also be liquefied under blast loading. Equations that relate peak compressive strain, peak pore pressure and residual pore pressure to explosive, distance, initial confining stress and relative density have been proposed, but the influence of relative density on residual excess pore pressure was found to be small, and blast-induced liquefaction can occur in dense sand. Magnitudes of particle velocity around 0.4cm/s were reported to have initiate blast-induced soil liquefaction, and past studies showed that multiple shocks significantly increase residual excess pore pressure.

1.2.3 Multi-hazard Engineering for Tunnel Lining

Transportation tunnels face the threats of multiple hazards in their life-cycle, which may include earthquake, fire, flooding, accidental impact, external explosion and internal blast, among others. These hazards' demands on structural capacity can be similar, coupled or contradicting, which impacts the selection of protection measures. For example, designing flexible barriers for protection against external explosion should consider the possible reduction of confinement on tunnel lining that may aggravate damage due to internal blast. Therefore, use of multi-hazard engineering is necessary in transportation tunnel protection, which requires comprehensive understanding of the responses of ground-tunnel system under all the hazard loadings considered. Although knowledge on the responses under earthquake loading and external blast may be available, the general interaction mechanism between tunnel and saturated soil under internal blast is lacking at present, which prevents the application of multi-hazard engineering in this context.

1.2.4 Centrifuge Modeling Technique

Centrifuge modeling is a proved approach for investigating geotechnical problems, including development of excess pore pressure, soil liquefaction and their effects on structure responses. For simulating explosion-related soil response and soil-structure interaction, it has been extensively employed since 1980's, the accuracy of which was demonstrated by the comparison of full-scale and centrifuge tests.

The centrifuge modeling of explosion is based on the scaling law of $W_{prototype} = W_{model} \cdot N^3$, in which $W_{prototype}$ is the weight of explosive in prototype scale, W_{model} is the one in model scale, and N is the centrifugal acceleration in g. That is to say, 1 gram of TNT under a centrifugal acceleration of $40g$ is equivalent to 64kg of TNT in prototype scale. Considering the threat facing transportation tunnels, very small amount of explosive is required to duplicate the effect of modest internal explosion in centrifuge. In most past studies, blasting cap or Exploded Bridgewire (EBW), generally used to ignite large amount of explosive, was used in geotechnical centrifuge to simulate prototype scale explosion.

Centrifuge modeling technique can also be employed in numerical simulation to reduce requirement of computing resources. Many examples can be found in the field of earthquake engineering. Tarek Abdoun presented results of eight centrifuge models of vertical single piles and pile groups subjected to earthquake-induced liquefaction and lateral spreading. Da Ha conducted four centrifuge tests designed to investigate the influence of pipe-fault orientation on pipe behavior under earthquake faulting.

However applications on the effects of blast loading on geotechnical structures are limited. Preece et al. employed the technique in their simulation of internal explosive loading on a model underground tunnel. Charlie conducted centrifuge model tests to investigate the response of medium dense sand due to blast. The influences of water content and saturation degree of medium dense sand were considered. Centrifuge models were set up at acceleration levels of $26.3g$ and $18.9g$ to simulate prototype tests in which 7kg TNT equivalent explosive was buried at a depth of 1.4m. The results from centrifuge model tests and the prototype tests matched well. De A studied the effects of surface blasts on underground structures through centrifuge model tests. 2.6mg of TNT equivalent explosives at an acceleration of $70g$ in centrifuge model tests simulated 0.9kg of TNT equivalent in prototype model. Different barrier systems were tested to study the mitigating effects.

1.2.5 Numerical Simulation of Explosion

Numerical simulation of explosion is also an established technique. Specifically, simulation of explosion in air and its effect on nearby structures has been possible by the development of multi-material Arbitrary Lagrangian-Eulerian (ALE) method. In this method, Lagrangian formulation is used to model solid continua and structures, and Eulerian formulation is used for large distortions of fluids, gases and explosives. The two domains are coupled through appropriate schemes.

Another method to simulate blast loading is to use CONWEP solution which has been embedded in the software LS-DYNA. This method is a purely Lagrange approach. Air blast pressure is computed with empirical equations (CONWEP) and directly applied to Lagrangian elements of the structure. A disadvantage of this approach is that reflections cannot be simulated when a blast encounters the interior corner of a room or where a structure is between the blast and the target structure.

In this study, a recently developed scheme for blast loading in LS-DYNA was used to investigate several numerical issues in the simulations of air blast and its effect on structures. In this scheme, a special layer of Eulerian elements is employed to receive the inflow of blast pressure from a point source of explosion. The blast pressure time-history at the ambient layer, viz. the pressure inflow, is determined using the CONWEP prediction. The blast pressure then propagates through the adjacent air domain to the targeted structure.

1.2.6 Numerical Simulation of Soil-Tunnel Interaction

Modeling of saturated ground-tunnel interaction is the other crucial aspect related to this study. State-of-the-art knowledge on structure damage, dynamic soil behavior, excess pore water pressure development, soil liquefaction and dynamic soil-structure interaction has resolved most issues related to this aspect.

One remaining problem is the description of blast wave propagation in saturated soils. Most numerical models so far have assumed undrained behavior of soils but theoretical analysis and numerical experiments showed that there exist two types of compression waves in permeable saturated ground subjected to blast loading, and that pore fluid and soil particle exhibit different velocities in these two waves.

Responses and damages of tunnel linings under internal blast loading are mainly governed by its interaction with the adjacent soil. Since volume-change is the key factor that determines compressive strain, residual excess pore pressure and tunnel response, sophisticated numerical procedure using FE program. LS-DYNA and assuming undrained soil response, which properly considers explosion simulation, structure damage, dynamic soil response, blast-induced pore pressure and dynamic soil-structure interaction, still has the potential to simulate saturated soil-tunnel interaction under internal blast loading.

As previously discussed, there are no validated design guidelines for transportation tunnels subjected to internal blast loading and the blast wave propagation inside the tunnel and in saturated soil is not well studied. This research aims to understand the soil-tunnel interaction due to medium blast loading with a series of numerical simulation in LS-DYNA. The magnitude of pressure and impulse applied on the tunnel due to blast will be studied and propagation process of blast wave inside the soil will be focused. After that the damage mode of tunnels and blast induced liquefaction of saturated soil will be understood. Proper mitigation measures will be investigated to improve tunnel safety considering multi-hazard environment.

1.3 Research Approaches

This study aims to revealing the knowledge on the interaction between circular subway tunnels and saturated soils subjected to internal explosion. It will also identify certain measure to mitigate structural damage. It will focus on the internal explosion with medium amount of explosive (<200kg of TNT-equivalent).

Numerical procedure using Finite Element program LS-DYNA that can model blast loading, air-solid interaction, material damage and soil-structure interaction will be used. A new coupling method will be used to study the process of blast wave propagation in the air domain inside the tunnel. This new coupling method can generate the blast wave accurately and apply the impact load on structures properly.

The excess pore water pressure inside the soil will be studied with an existing soil model (FHWA) which can simulate pore water pressure and effective soil pressure. Thin layer interface elements will be used to describe the transitive domain between tunnel and soil. An isotropic elastic-plastic material in LS-DYNA will be used to simulate the cast-iron tunnel lining in the numerical analysis. In this model yield stress versus plastic strain curves can be defined for compression and tension. Thus this model can simulate the cast-iron whose compressive strength is much larger than tensile strength.

Centrifuge modeling technique will be a research tool in this study. Centrifuge modeling will save computer resources and offer simplified but accurate simulation of full scale geotechnical structures subjected to blast loading as shown in Figure 1.4.

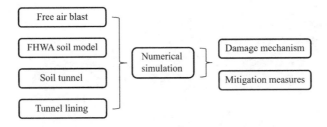

Figure 1.4 Illustration of research approach

1.4 Chapter Contents

This thesis is subdivided into six specific chapters dealing with various aspects of the research. The titles of these chapters are as follows:

Chapter 1 Introduction and Literature Review

Chapter 2 Blast Simulation Involving Air-Structure Interaction

Chapter 3 Numerical Modeling of Soil-Tunnel Interaction
Chapter 4 Response of Circular Tunnels Subjected to Medium Internal Blast Loading
Chapter 5 Parametric Study
Chapter 6 Conclusions and Future Research

Chapter 2 Blast Simulation Involving Air-Structure Interaction

2.1 Blast Wave Propagation in Air

2.1.1 Theoretical Background

Blast wave propagation in air is governed by three fundamental equations in fluid dynamics: mass conservation, momentum conservation and energy conservation equations, which are respectively listed in the following.

Mass conservation equation:

$$\frac{\partial \rho}{\partial t} + \frac{\partial}{\partial x_i}(\rho v_i) = 0 \tag{2.1}$$

Momentum conservation equation:

$$\frac{\partial}{\partial t}(\rho v_i) + \frac{\partial}{\partial x_j}[(\rho v_i)v_j] - \frac{\partial}{\partial x_j}(\sigma_{ij}) - \rho a_i = 0 \tag{2.2}$$

Energy conservation equation:

$$\frac{\partial(\rho E)}{\partial t} + \frac{\partial}{\partial x_i}(\rho v_i E) - \frac{\partial}{\partial x_i}\left(k\frac{\partial T}{\partial x_i}\right) + \frac{\partial}{\partial x_i}(p v_i) - \frac{\partial}{\partial x_i}(s_{ij} v_j) - \rho a_i v_i - q_H = 0 \tag{2.3}$$

Here, ρ is the fluid density, v_i is the fluid velocity, σ_{ij} is the fluid stress, s_{ij} is the deviatoric stress, T is the fluid temperature, k is the coefficient of thermal conductivity, p is the fluid pressure, a_i is the fluid acceleration, E is the fluid total energy [$E = \Im_i(T, p) + \frac{1}{2} v_i v_i$, with $\Im_i$ being the intrinsic energy per unit mass], and q_H is the heat associated with chemical reaction in the fluid.

These three equations can be combined to yield the Navier-Stokes equation:

$$\frac{\partial \boldsymbol{\Phi}}{\partial t} + \frac{\partial \boldsymbol{F}_i}{\partial x_i} + \frac{\partial \boldsymbol{G}_i}{\partial x_i} + \boldsymbol{Q} = 0 \tag{2.4}$$

In which

$$\boldsymbol{\Phi} = [\rho, \rho v_1, \rho v_2, \rho v_3, \rho E]^{\mathrm{T}}$$

$$\boldsymbol{F}_i = [\rho v_i, \rho v_1 v_i + p\delta_{1i}, \rho v_2 v_i + p\delta_{2i}, \rho v_3 v_i + p\delta_{3i}, \rho H v_i]^{\mathrm{T}}$$

$$\boldsymbol{G}_i = \left[0, -s_{1i}, -s_{2i}, -s_{3i}, -s_{ij}v_j - k\frac{\partial T}{\partial x_i}\right]^{\mathrm{T}}$$

$$\boldsymbol{Q} = [0, \rho a_1, \rho a_2, \rho a_3, \rho a_i v_i - q_H]^{\mathrm{T}}$$

Here, δ_{ij} is the Kronecker delta, and H is the enthalpy of the fluid.

Analytical solution of the Navier-Stokes equation is a formidable task. Imposing simple boundary condition, simple geometry, simple equation of state for air and simple energy equation for air, Neumann gave the solution of blast for a point source explosion. He assumed that the equation of state for air is $p=c\rho T$, and the total energy for air is $E=[c/(\gamma-1)]T$, with c being the specific gas constant and defined as $c=c_p-c_v$, and γ being the heat capacity ratio and defined as $\gamma=c_p/c_v$. Here c_p is the specific heat under constant pressure and c_v is the one under constant volume. He showed that the blast pressure in the air can be expressed as:

$$p = \frac{8}{25(\gamma+1)}\rho_0 f(t,\gamma,\theta) \qquad (2.5)$$

Here ρ_0 is the initial density of the air before blast, t is time, and θ is a parameter representing the location of the point of interest in the free air domain. $f(t, \gamma, \theta)$ is a complicated function of t, γ and θ, the detail of which can be found in Neumann. This solution is valid for $1<\gamma<2$.

Equation (2.5) showed that, by increasing the initial density ρ_0, the blast pressure in the air due to explosion would increase. This theoretical fact can be employed to increase the blast pressure in numerical simulation and to partially compensate the loss of energy due to large element size.

2.1.2 Blast Wavelength and Finite Element Size

Element-based simulation of wave propagation in solid or fluid continuum is closely related to element size and wavelength. Adequate number of elements per wavelength is required to capture the wave propagation, which is also related to the element type and element shape. In the simulation of linear time-harmonic acoustics, some studies showed that around 10 linear elements per wavelength was necessary to maintain an error that was less than 10%. However, this conclusion cannot be directly applied to blast wave propagation problems, particularly those involving air-structure interaction, as the shape of a blast wave is much more complicated than that of a harmonic acoustic wave. It is expected that more linear elements would be needed.

There exist a few studies investigating the required element size in the simulations of blast waves and their effects on structures. Chapman et al. found that a grid size less than 3mm was necessary to simulate the blast pressure and specific impulse at a scaled distance of 0.95m/kg$^{1/3}$. The amount of explosive simulated was 75g TNT. Luccioni et al. compared their simulated peak pressures with those presented in Kinney and Graham, and concluded that a grid size of 100mm was adequate to simulate the peak pressure due to the explosion of 100kg TNT. The parametric study by Shi et al. showed that a grid size of 10mm was necessary to capture the peak pressure and specific impulse at a scaled distance of 1.0m/kg$^{1/3}$ due to an explosion of 1000kg TNT. These analyses were all carried out using the Finite Element code AUTODYN.

These previous studies showed that the required element size to satisfy adequate accuracy depended both on the amount of explosive and on the scaled distance. Generally, larger element size may be used if large-scale explosion is simulated, and the strict requirement of element size relaxes if the scaled distance of interest is large. However, it is still difficult to draw a conclusion on the element size, amount of explosive and scaled distance based on available literature.

The relationship between element size and amount of explosive is related to blast wavelength. According to UFC 3-340-02, there exists a unique scaled wavelength (unit: $L_w/W^{1/3}$) of incident blast due to free air explosion. This means that, larger amount of explosive resulted in larger wavelength, which may in turn relax the requirement on element size. However, it is not known whether the same scale ratio can be used to determine the element size for large-scale explosion simulation based on that of a small-scale explosion simulation. For example, if a grid size of 3mm is adequate to simulate the blast wave propagation due to 100g TNT, it is not known whether a grid size of 30mm is enough for an explosion of 100kg TNT. The problem would be more complicated in fluid-structure interaction due to the complexity induced by wave reflection and wave superposition.

2.2 Simulation of Blast Loading in LS-DYNA

2.2.1 Modeling of Air and Explosives

1. Material models

The air and explosives in all the numerical simulations were modeled using the 3D solid elements in LS-DYNA. The elements were 8-noded linear hexahedron elements integrated using 8-point Gaussian method.

The air was assumed to be ideal gas and modeled using the MAT _ NULL material model. Null material has no shear stiffness and hourglass control must be used with great care. In some applications, the default hourglass coefficient in LS-DYNA might lead to significant energy losses. The hourglass control scheme that was used by Schwer to successfully model ideal gas was employed in this study.

In the MM-ALE approach the explosive was modeled with *MAT _ HIGH _ EXPLOSIVE _ BURN. It allows the modeling of the detonation of a high explosive. In addition an equation of state must be defined.

Burn fractions, F, which multiply the equations of states for high explosives, control the release of chemical energy for simulating detonations. At any time, the pressure in a high explosive element is given by:

$$p = F p_{eos}(V, E) \qquad (2.6)$$

where p_{eos}, is the pressure from the equation of state, V is the relative volume, and E is the internal energy density per unit initial volume. The burn fraction F is taken as the

maximum:
$$F = \max(F_1, F_2) \tag{2.7}$$
Where
$$F_1 = \begin{cases} \dfrac{2(t-t_1)DA_{e\max}}{3v_e} & \text{if } t > t_1 \\ 0 & \text{if } t \leqslant t_1 \end{cases} \tag{2.8}$$

$$F_2 = \dfrac{1-V}{1-V_{CJ}}$$

where V_{CJ} is the Chapman-Jouguet relative volume and t is current time. D is the detonation velocity and t_1 is a lighting time. If F exceeds 1, it is reset to 1. This calculation of the burn fraction usually requires several time steps for F to reach unity, thereby spreading the burn front over several elements. After reaching unity, F is held constant. This burn fraction calculation is based on work by Wilkins (1964) and is also discussed by Giroux (1973).

2. Equation of states

An equation relating the pressure, temperature, and specific volume of a substance is defined as an equation of state. Two equations of state were used to simulate the air: EOS _ LINEAR _ POLYNOMIAL and EOS _ GRUNEISEN.

In the linear polynomial state form the pressure is expressed as following:

$$p = (\gamma - 1)\dfrac{\rho}{\rho_0}E_0 \tag{2.9}$$

Here γ is the ratio of specific heat and was assumed as 1.4 according to previous numerical experiences. ρ_0 is the initial density of air and was assumed to be $0.00129 g/cm^3$. This value was increased in some of the analyses to investigate its effect. E_0 was given a value of 0.25MPa according to Schwer and assuming an initial air temperature of 20 ℃.

In the Gruneisen state form the pressure is expressed as following:

$$p = \rho_0 C^2 \mu + (\gamma_0 + \alpha\mu)E \tag{2.10}$$

Where μ is the relative volume which is defined as

$$\mu = \dfrac{1}{V} - 1 \tag{2.11}$$

E is the internal energy per initial volume and C is the intercept of the u_s-u_p curve. γ_0 and α are user defined input parameters. ρ_0 was assumed to be $0.00129 g/cm^3$ similar as above. E_0 was given a value of 0.25MPa as before. γ_0 is the ratio of specific heat and was assumed as 1.4. C was assumed to be 0.344 according to numerical experiences. Here α used the default value in the LS-DYNA.

The JWL equation of state is used for detonation products of high explosives. The pressure is defined as

$$p = A\left(1 - \dfrac{\omega}{R_1 V}\right)e^{-R_1 V} + B\left(1 - \dfrac{\omega}{R_2 V}\right)e^{-R_2 V} + \dfrac{\omega E}{V} \tag{2.12}$$

The parameters ω, A, B, R_1 and R_2 are constants pertaining to the explosive. This EOS is well suited because it determines the explosive's detonation pressure in applications involving structural metal accelerations. In the analysis the parameter A was assumed to 3.71 and B was assumed to 0.0743. R_1 and R_2 were set up to 4.15 and 0.95 respectively. ω was assumed to be 0.3 and the initial internal energy per initial volume was assumed to be 7GPa.

2.2.2 MM-ALE Approach

Traditionally blast loading was simulated by employing the multi-material Arbitrary Lagrange Eulerian solver (MM_ALE). In this approach, Lagrangian formulation was used to model solid continua and structures, and Eulerian formulation was used for large distortions of fluids, gases and explosives. The two domains were coupled through appropriate schemes. The explosive must be properly modeled and a large number of fluid elements must be simulated between the explosive and structures to transfer the blast wave as shown in Figure 2.1. Previous study showed that very small element size (about 0.5mm) is necessary to properly model the explosive and to accurately simulate the blast pressure released by the explosive. This requirement of element size, together with the large number of fluid elements required between the explosive and the structure, makes it extremely demanding to complete a full-scale simulation involving large structures.

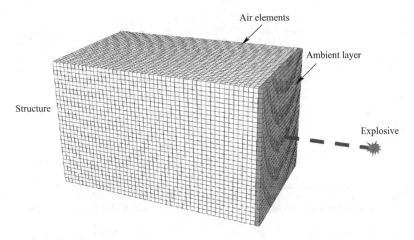

Figure 2.1 Illustration of the Load _ Blast _ Enhanced approach in LS-DYNA

A series of simulations were conducted to gain the proper parameters of explosive model and air model. Two different equations of air were compared: EOS _ LINEAR _ POLYNOMIAL and EOS _ GRUNEISEN. Other parameters in the material model MAT _ NULL were validated. Simulation results are shown in Figure 2.2 to Figure 2.7. Main observations are: EOS _ LINEAR _ POLYNOMIAL behaved overall better than EOS _ GRUNEISEN. It is also found that increasing air temperature T and adjusting specific heat ratio r might result in better comparison of air pressure with CONWEP.

In this set of simulation, the size of air element was 40mm, while that of the explosive element was 10mm. With these parameters, the results of air pressure compared poorly with the CONWEP prediction, particularly when the scaled distance was small. Much smaller element size has to be employed in order to obtain better accuracy. This is quite beyond the capacity of our existing computer resources. This approach is therefore abandoned in this study.

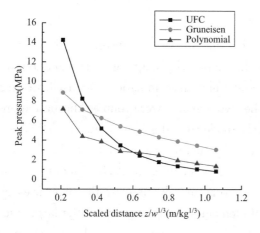

Figure 2.2 Peak pressures with different EOS (TNT 104.32kg)

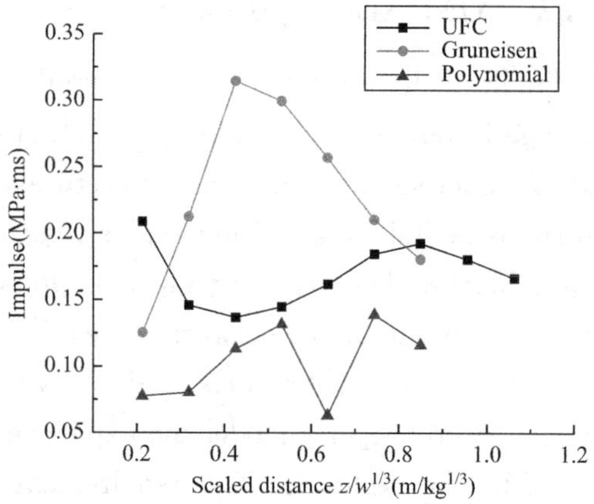

Figure 2.3 Incident impulses with different EOS (TNT 104.32kg)

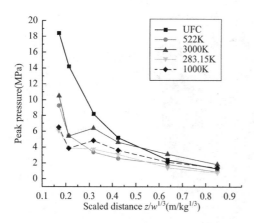

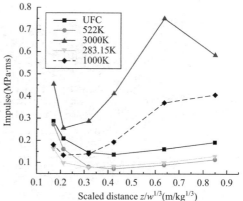

Figure 2.4 Peak pressures with different T (TNT 13.04kg)

Figure 2.5 Incident impulses different T (TNT 13.04kg)

2.2.3 LBE+MM-ALE Approach

There also exists a simplified approach, which imposes pressure time-history on structure surfaces based on the CONWEP reflected pressure on a rigid surface. This approach significantly reduced the number of elements and has been used in many blast designs of structures. A disadvantage of this approach is that wave reflection and superposition cannot be accounted for as would occur in the interior corner of a structure or inside a tunnel. The reflected pressure time-history on a deformable surface is also not the same as that on a rigid surface, and Borvik et al. have demonstrated that this approach resulted in large errors even in the simulation of a simple container structure subjected to external blast loading.

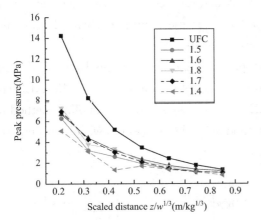

Figure 2.6 Peak pressures with different r (TNT 104.32kg)

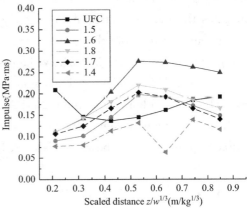

Figure 2.7 Incident impulses with different r (TNT104.32kg)

The combination of the CONWEP incident pressure and the MM-ALE solver can be used to overcome some of the disadvantages and to obtain better results with less computational efforts. This coupling method is explained using an example of internal explosion inside a tunnel, as shown in Figure 2.1. Only the air immediately surrounding the target structure is modeled with an ALE domain. Blast pressures and time-histories are applied to a layer of ALE elements (the ambient layer is shown in Figure 2.1), which faces the explosive charge and acts as a source for the adjoining air elements. The blast wave then propagates through the air domain and eventually interacts with the structure. The loading on the structure due to wave reflection and superposition can then be captured by the ALE domain. This capacity can be realized in LS-DYNA by employing the key word LOAD_BLAST_ENHANCED and defining a layer of special elements (the ambient layer). Figure 2.8 compares the blast-pressure inside the ambient layer with that of CONWEP in a blast simulation with a scaled distance $2.62 \text{m/kg}^{1/3}$. Almost identical results were obtained.

It is noted that similar approaches have been used in blast simulations in previous studies. Similar to the LOAD_BLAST_ENHANCED in LS-DYNA, the "inflow" of blast pressure was realized by defining a special layer of Eulerian Finite Elements that have properties different from those of the adjacent air elements, and by employing some empirical or semi-empirical formulae of incident blast pressure.

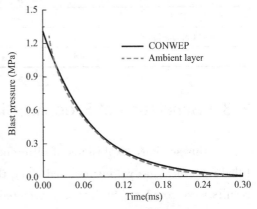

Figure 2.8 Blast pressure inside ambient layer with CONWEP solution

Another approach was proposed to simulate loads on bridge elements similar to that by CONWEP (Yi zhihua). In this approach, CONWEP pressure generated for a specific charge weight is transferred to an air layer near structural element. The blast wave front propagates through the surrounding air layer. The air mesh interacts with Lagrangian structure element to apply the load on structural elements.

Figure 2.9 and Figure 2.10 show the peak pressure and incident impulse in air domain due to 104.32g TNT equivalent explosion with new coupling approach and MM-ALE approach. It took 2.5 hours to run one MM-ALE case and just 1 hour for coupling method using the same computer. The new coupling method led to better results with less computer resource requirements as shown in Table 2.1.

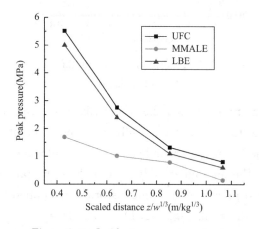

Figure 2.9 Incident overpressure in air domain (TNT 104.32g)

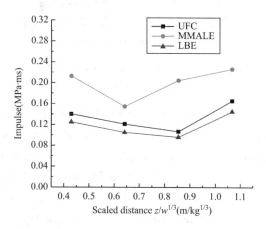

Figure 2.10 Incident impulse in air domain (TNT 104.32g)

Results comparison of two simulation methods   Table 2.1

	MM-ALE	New coupling
Air domain	30cm×30cm×30cm	50cm×30cm×15cm
Mesh Size	Air 4mm; explosive 1mm	Air: 4mm
Element No	421875	393300
Calculation time	2.5hours	1.0hours
Accuracy	40%~50%	8%~12%

2.3 Simulation of Free Air Blast

In this section, explosions of different amounts of spherical TNT in an infinite space filled with ideal gas were simulated using the new blast loading scheme in LS-DYNA. The results were compared with those predicted by CONWEP. In all of the analysis, the ambient layer was a 0.4cm thick layer of element standing between the block of air elements and the point source of explosion.

2.3.1 Effects of Initial Air Density

In this set of analysis, the length, width and height of the hexahedron elements were all 4mm. The air domain was simulated as a rectangular block of air elements, the dimension of which was 14.6cm×30cm×55cm. A spherical TNT of 8kg was assumed to locate 30.2cm (scaled distance=0.151m/kg$^{1/3}$) away from the ambient layer. Three initial air densities were tested: the original air density ρ_0=0.00129g/cm^3, 1.2 of the original value, and 1.4 of the original value, the latter two of which corresponds to air density factors of f_ρ=1.2 and f_ρ=1.4.

Figure 2.11 and Figure 2.12 show the relationship between scaled distance and incident peak blast pressures and positive specific impulses. It can be seen that with an increase in the initial air density, both the simulated peak blast pressure and specific impulse increased. Compared with the CONWEP predictions, an air density factor of 1.2 yielded the best results with a grid size of 4mm. The numerical simulation was not as good at small scaled distance (< 0.2m/kg$^{1/3}$), although the values in the ambient layer was very close to those predicted by CONWEP. Similar results were also reported by other investigators in their simulation of free air blast.

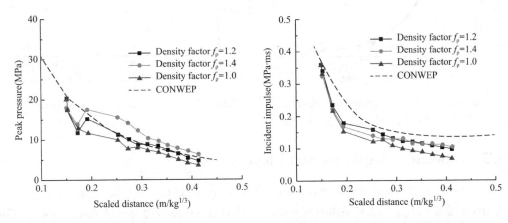

Figure 2.11 Peak pressure with different density factor at different scaled distances

Figure 2.12 Impulse with different density factor at different scaled distances

It is noted that, with 8kg TNT, the numerical simulations with a density factor of 1.2 were very close to the CONWEP predictions for the range of scaled distance around 0.4 m/kg$^{1/3}$. This is very relevant for the simulation of medium-scale explosion and its effect on close-by structures.

2.3.2 Simulation of Free Air Blast with Different Locations of Ambient Layer

The new blast loading scheme provides a mechanism to optimize the locations of ambient layer to the source of explosion, provided that the thickness of air domain over the structure surface is adequate to capture the wave reflection and superposition. In this series of analyses, the same element size of 4mm was used with an air density factor of 1.2. The block of air elements had a dimension of 14.6cm×30cm×55cm. With different distances of detonation from the ambient layer and different amounts of explosive, the peak incident pressure and specific impulse inside the air domain were calculated and compared with the CONWEP predictions. The amount of explosives ranged from 88.14g to 352.56g.

Figure 2.13 and Figure 2.14 show the comparisons of peak incident pressures and specific impulses. It can be seen that it is possible to choose an optimal distance between the source of explosion and ambient layer such that the peak incident pressure was almost accurate while the specific impulse was very satisfactory, for a range of scaled distance between 0.4m/kg$^{1/3}$ and 2.0m/kg$^{1/3}$.

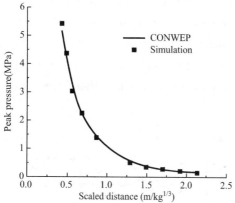
Figure 2.13 Peak incident pressures and specific impulses for different scaled distance

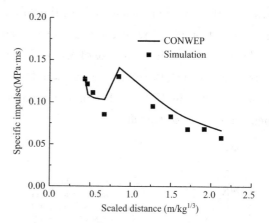
Figure 2.14 Peak incident pressures and specific impulses for different scaled distance

2.3.3 Amount of Explosive and Element Size

The required element size to capture blast wave propagation is related to the amount of explosive, as a unique relationship between scaled distance and wavelength exists. This fact is demonstrated in Figure 2.15 and Figure 2.16, which shows the numerical simulations of three free air blasts. The amounts of TNT employed were 0.025g, 1.6kg and 8.0kg, respectively; the element size was still 4mm; and the air density factor was 1.2. Apparently, the case with 0.025kg TNT was the worst, while the case with 8.0kg of TNT yielded the best results in terms of incident peak pressure and specific impulse. The case with 1.6kg of explosive was also satisfactory. The results show that smaller element size is needed while simulating small-scale explosion.

However, proportional scale of element size based on the wavelength did not seem to work in the numerical exercises. Figure 2.17 and Figure 2.18 compare the numerical simulations with 1.2kg TNT and 200kg TNT, one with 4mm element size and the other with 22mm element size. The scaled element size by the weight of explosive was basically the same in these two cases. However, the 200kg case resulted in much smaller incident peak pressure and specific impulse. Larger air density factors were then used in the 200kg case, but the resulted specific impulses were not satisfactory, as shown in Figure 2.18, although the simulated incident peak pressures improved.

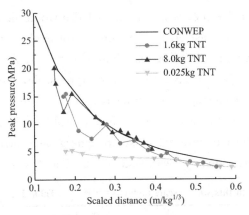

Figure 2.15 Incident pressures of different amount of explosive

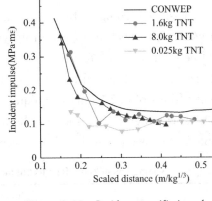

Figure 2.16 Incident specific impulses of different amount of explosive

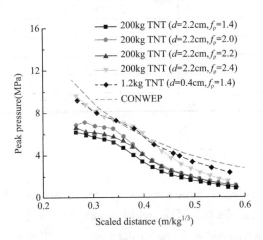

Figure 2.17 Incident pressures of different amount of explosive with different mesh size and air density

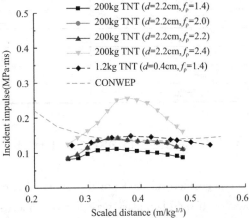

Figure 2.18 Incident specific impulses of different amount of explosive with different mesh size and air density

The results in Figure 2.17 and Figure 2.18 showed that, for relatively small standoff distance that corresponds to a scaled distance smaller than $1.0 m/kg^{1/3}$, the numerical simulation would be very demanding if relatively large amount of explosive is used. For example, for an explosion of 200kg TNT and a standoff distance of 3m, in order to properly simulate the specific impulse, a very large number of air elements have to be used, and the element size probably should be around 10mm. The computation effort would be very large even if the new blast loading scheme is used.

2.4 Centrifuge Technique

One alternative is to use the centrifuge technique that has been successfully employed in experiments to simulate the effects of blast. Using this technique, the gravity acceleration would be increased to n times of the original value, while the model size would be reduced by n times. According to the scaling law of blast effect, the amount of explosive in the model would then be $1/n^3$ of the prototype amount. For example, if $n=5$, 1.6kg of TNT could be used to simulate the effect of 200kg of TNT in the prototype scale, while the model size could be reduced by 5 times. The element size can be as large as 4mm using an air density factor of 1.2. This technique is particularly relevant in the problems involving soil structure interaction, such as the effect of internal blast on underground structures. Table 2.2 shows the scaling law of centrifuge tests with a scale factor 5 in this study.

Scaling law of centrifuge technique in this study (scale factor $N=5$)　　　Table 2.2

	Prototype	Simulation
Acceleration	g	$5g$
Length	L	$L/5$
TNT amount	W	$W/125$
Displacement	Δ	$\Delta/5$
Stress	σ	σ
Strain	ε	ε
Time	t	$t/5$
Frequency	f	$5f$

2.5 Reflected Blast on a Rigid Surface

The geometrical model of this series of simulations was the same as the one in section 2.3. However, a rigid boundary was placed inside the air domain. The rigid boundary is model with one layer of extremely stiff solid element which overlapped with the air domain. Five different amounts of TNT, 88.1g, 176.3g, 264.4g, 352.6g and 8000g, were tested, the scaled distances of which to the rigid boundary ranged from 0.35m/kg$^{1/3}$ to 0.68m/kg$^{1/3}$. The simulations of free air blast with the same amounts of TNT were also carried out and the results at the interested scaled distances were very close to those predicted by CONWEP.

Figure 2.19 and Figure 2.20 show the results and their comparisons with the CONWEP predictions. Except the reflected specific impulse at 0.35m/kg$^{1/3}$ scaled distance, the other simulation results were very close to the predictions by CONWEP. The reflected specific impulse at 0.35m/kg$^{1/3}$ scaled distance was about 20% smaller than the CONWEP prediction. The smaller reflected specific impulse at this location could be due to the small scaled distance.

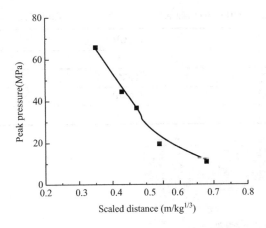 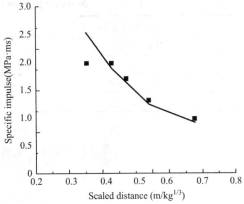

Figure 2.19 Reflected pressures at different scaled distances

Figure 2.20 Reflected specific impulses at different scaled distances

2.6 Simulation of Blast Effect on a Steel Plate

Boyd carried out two identical experiments (E14 and E15) to investigate the effect of air blast on a steel plate. Figure 2.21 shows the experiment setup. A 5mm thick steel plate was supported by four concrete blocks. 250g of Pentolite explosive was denoted 50cm above the steel plate. The steel plate was attached to a steel ring and placed directly on the concrete blocks. The steel plate was instrumented for vertical displacement at the center, accelerations at two locations, and overpressures at another two locations, as shown in Figure2.21 (b).

Due to symmetry, only half of the system was modeled in LS-DYNA as shown in Figure 2.22. The air domain above the steel plate was modeled by two approaches, one with 10mm element size and an air density factor of 1.0, and the other with 15mm element size and an air density factor of 1.2. Larger elements could be used herein as the scaled distance of interest was larger than $0.8 m/kg^{1/3}$, which were selected by comparing the incident air pressure from numerical simulation to that by UFC 3-340-02 (2008). The thickness of air domain above the steel plate was 255mm, and on its top laid the ambient layer. The steel plate was meshed with Lagrangian solid elements and coupled with the Eulerian elements of the air. In order to capture the bending response, 5 layers of solid elements were used along the thickness of the steel plate. The steel plate was modeled by the isotropic elasto-plastic model, with a Young's modulus of 203GPa, a Poinsson's ratio of 0.3, and a yield strength of 270MPa according to Boyd (2000). The steel ring was simulated by elastic material, with a Young's modulus 10 times larger than that of the steel plate. Frictional surface-to-surface contact was applied between concrete blocks and the steel ring. 250g Pentolite explosive was converted into equivalent TNT explosive and used in the analysis (Boyd, 2000).

Table 2.3 Material parameters for the steel plate

Parameter	Value
Young's Modulus, E	203GPa
Poissons Ratio v	0.3
Yield Stress σ_0	270MPa
Tangent Modulus E_T	470MPa
Density ρ	7.85g/cm^3

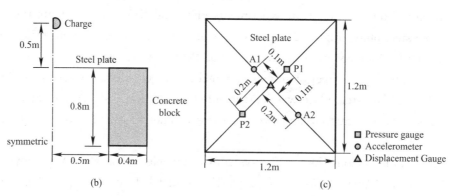

Figure 2.21 Explosion experiment for a steel plate (Boyd, 2000)

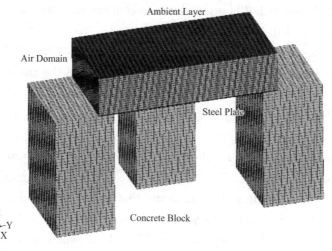

Figure 2.22 LS-DYNA simulation model of explosion experiment

The results of both simulations matched well with the experimental data, as shown in Table 2.4. The peak reflected pressures simulated were smaller than the measured ones. However Boyd (2000) reported that the traces from the pressure transducers exhibited a large amount of noise at the peak. The measured overpressure jumped and reduced immediately, causing very small difference in the specific impulse, which indicated that it could be due to some noise in the measurement.

Comparisons of tested and simulated results   Table 2.4

Results of simulation	E14	E15	I	II
Displacement at the plate center (mm)	−33.0		−32.4	−36
Acceleration at position A1 (g)	14657	13185	12609	12884
Acceleration at position A2 (g)	14768	14239	16709	13560
Overpressure peak at P1 (MPa)	9.4	9.4	5.9	4.9
Overpressure peak at P2 (MPa)	8.7	8.0	4.9	4.63
Impulse at position P1 (MPa·ms)	0.44	0.38	0.44	0.46
Impulse at position P2 (MPa·ms)	0.38	0.33	0.36	0.52

Note: Simulation I used 10mm element size and the original initial air density; simulation II used 15mm element size and 1.2 times of the original initial air density.

2.7 Size of Air Domain Inside the Tunnel

A proper thickness of air domain between ambient layer and tunnel was evaluated in order to properly simulate blast wave propagation inside a circular tunnel. The size of air domain was firstly determined by modeling the explosion inside a rigid tunnel. The air element size was 4mm while the air density ratio was 1.2. With a scale factor of $N=5$, the 5m diameter tunnel in prototype scale was scaled down to 1m diameter tunnel. One eighth models were used to save computer resource due to symmetry. Six different air domain thicknesses were tested here: 38%, 30%, 24% and 20% of the tunnel radius. The length of tunnel in the 38% case and 30% fine case was 15m in prototype scale. However, from the results it was found that only the blast wave pressure in the first third domain of tunnel was significant. Therefore only the first third domain was modeled in the 30% coarse case, 24% case and 20% case. The length of the tunnel in these cases was 5m in prototype model. The details of the models are shown in Table 2.5.

The blast wave propagations from 50kg TNT equivalent and 200kg TNT equivalent in prototype scale were evaluated. The peak pressure and specific impulse are shown in Figure 2.23~Figure 2.26. It can be seen that the peak pressure was not considerably affected by the thickness of air domain and it was close to the peak reflected pressure on a plane rigid surface from UFC 3-340-02 (2008), but the specific impulse was related to the air thickness. When the air thickness was larger than 30% of the tunnel radius, the difference was very small. With an air thickness of 24% tunnel radius, the specific impulse was some larger,

Trial simulations to determine proper air domain thickness   **Table 2.5**

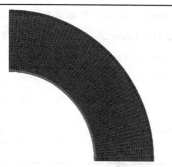 Air thickness=38% of tunnel radius Mesh size 4mm element number 379900 TNTequivalent: 200kg, 50kg	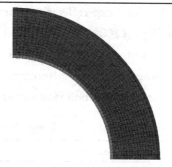 Air thickness=30% of tunnel radius Two cases: the fine one mesh size is 4mm and element number is 244800. coarse one is the same as 6cm case and element number is 64800 TNT equivalent: 200kg, 50kg

 Air thickness=24% of tunnel radius
 	
mesh size in transverse direction 4mm, axial direction half is 5mm and half is 1cm
  element number 50400
  TNT equivalent: 200kg, 50kg
	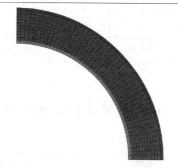 Air thickness=20% of tunnel radius
	
	mesh size is the same as 6cm case
	  element number43200
	  TNT equivalent: 200kg, 50kg

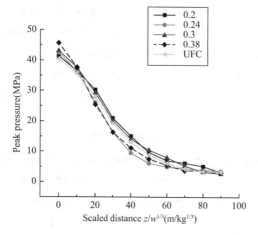

Figure 2.23 Peak pressure in air next to tunnel ($N=5$, 200kg TNT $D=5$m)

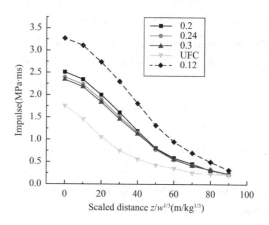

Figure 2.24 Impulses in air next to tunnel ($N=5$, 200kg TNT $D=5$m)

and it might result in conservative response of subway tunnel under blast loading. The specific impulse with sufficient thickness of air domain was higher than the peak reflected one on a plane rigid surface due to effect of blast wave reflection and superposition in the tunnel. In addition, along the axis of the tunnel, the peak pressure and specific impulse were very small when the incident angle on the lining was 75°.

2.8 Conclusions and Discussions

In this chapter, a recently developed scheme for blast loading in LS-DYNA was used to investigate several numerical issues in the Finite Element simulation of blast loading inside a circular tunnel. In this scheme, a special layer of Eulerian elements is employed to receive the inflow of blast pressure from a point source of explosion. The blast pressure time-history at the ambient layer, viz. the pressure inflow, is determined using the CONWEP prediction. The blast pressure then propagates through the adjacent air domain to the targeted structure.

The objective of this chapter is to validate this blast loading scheme, to investigate the combined effect of element size and initial air density, and to shed lights on the simulation of medium-scale explosion (50~200kg TNT) inside a circular tunnel.

The following conclusions were obtained from this chapter.

(1) The new blast loading scheme could be used to properly simulate air blast, provided that proper element size corresponding to a specific amount of explosive is used. This blast loading scheme can significantly reduce the number of air elements and totally eliminate the explosive elements, thus lowering the computational effort.

(2) The initial density of air in the numerical model could be purposely increased when the equation of state for ideal gas is used to model the air domain. The increase of air density could partially compensate the error induced by the use of relatively large air elements.

(3) Although relatively larger element size could be used in the modeling of large air blast, the required element size was not proportional to the cubic-root of the weight of explosive W, although there exists a unique wavelength $L_w/W^{1/3}$ as a function of scaled distance according to UFC 3-340-02 (2008). Centrifuge technique, which has been successfully employed in model tests to investigate the blast effects, may be used when simulating medium- to large scale explosion.

(4) As long as the free air blast at a specific scaled distance was properly simulated, the fluid structure interaction at the same location could be properly duplicated using proper Arbitrary Lagrangian Eulerian (ALE) coupling scheme.

(5) The size of air domain can be significantly reduced in the 3D Finite Element simulation of internal blast in a circular tunnel. The proposed approach resulted in more realistic and more severe failure in the tunnel lining when compared with the direct application of blast pressure on the lining surface from CONWEP.

It was also found that at small scaled distance ($< 0.3\text{m/kg}^{1/3}$), the new blast loading scheme also resulted in smaller peak pressure and specific impulse, even if they were accurate in the boundary layer of inflow pressure (the ambient layer). This is consistent with previous simulations that modeled the explosive and the surrounding air (Luccioni et al., 2006; Shi et al., 2011). However, compared to the conventional approach, the new blast loading scheme has the potential to duplicate the peak blast pressure and specific impulse at a scaled distance around $0.4\text{m/kg}^{1/3}$. This is very relevant for practical applications.

Finally, it should be pointed out that the new blast loading scheme cannot simulate the effects of fragments from some sources of explosions.

Chapter 3 Numerical Modeling of Soil-Tunnel Interaction

3.1 Modeling of Tunnel Lining

An isotropic elastic-plastic material in LS-DYNA was used to simulate the cast-iron tunnel lining in the numerical analysis. In this model yield stress versus plastic strain curves can be defined for compression and tension. Thus this model can simulate the cast-iron whose compressive strength is much larger than tensile strength.

The stress-strain behavior follows one curve in compression and another in tension. The sign of the mean stress determines the state wherea positive mean stress (i. e. , a negative pressure) is indicative of tension. Two load curves, $f_t(p)$ an $f_c(p)$ are defined, which give the yield stress σ_y, versus effective plastic strain for both the tension and compression regimes. The two pressure values p_t and p_c when exceeded, determine if the tension curve or the compression curve is followed, respectively. If the pressure p falls between these two values, a weighted average of the two curves is used:

If
$$-p_t \leqslant p \leqslant p_c$$

$$\begin{cases} scale = \dfrac{p_c - p}{p_c + p_t} \\ \sigma_y - scale \cdot f_t(p) + (1 - scale) \cdot f_c(p) \end{cases} \quad (3.1)$$

Strain rate is accounted for using the Cowper and Symonds model, which scales the yield stress with the factor:

$$1 + \left(\dfrac{\dot{\varepsilon}}{C}\right)^{\frac{1}{p}} \quad (3.2)$$

where $\dot{\varepsilon}$ is the strain rate $\dot{\varepsilon} = \sqrt{\dot{\varepsilon}_{ij}\dot{\varepsilon}_{ij}}$.

Figure 3.1 and Figure 3.2 shows the stress and strain relations in tension and compression.

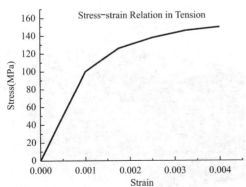

Figure 3.1 Stress strain relation of tunnel lining in tensile direction

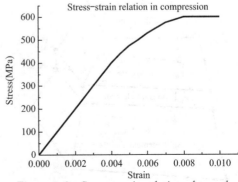

Figure 3.2 Stress strain relation of tunnel lining in compressive direction

The Young's modulus of cast iron tunnel is 100GPa. In the tensile direction, the tunnel is in the elastic domain before the strain reaches 0.001. The failure stress is 150MPa when the tensile strain equals to 0.004. In the compressive direction the yield stress of tunnel is 600MPa.

In the numerical analysis failure flag was set up to plastic strain limit control. When the plastic strain reaches the limit value, the element is deleted from the calculation.

3.2 Modeling of Soil Tunnel Interfaces

Thin-layer elements were used to simulate the interface between soil and tunnel. The concept was proposed by Desai (65). The element thickness is much smaller than the other dimension (2D) or the other two dimensions (3D). Ordinary solid elements are still used to simulate the interfaces just the thickness is about 0.01~0.1 of the longer dimension. Figure 3.3 illustrates the application of thin-layer elements.

Mohr-Coulomb elasto-plastic model is employed to simulate the interface behavior. The parameters in the constitutive model should accord well with physical characteristics from experiments. Generally the strength of the thin interface is smaller than soil strength while the normal stiffness is larger than that of soil. The shear modulus may be similar to that soil.

The main advantage of thin layer interface is no interface penetration under large compressive loading, which is more numerically stable. Thin-layer elements can also model dilation with a proper constitutive model. But thin-layer elements may not be suitable for large deformation analysis.

A specific case was simulated in LS-DYNA to check the validity of interface between soil and tunnel as shown in Figure 3.4. An impact load with similar magnitude and duration with blast load was applied on the tunnel and the interface vibrates with tunnel. As shown in Figure 3.5 there is no tensile stress in the interface when there is tensile strain. It indicates that the approach can reproduce the separation of soil-tunnel interface during lining vibration due to internal blast loading.

In this study, the strength of soil-lining interface was assumed to be 2/3 of soil strength, while its stiffness and dilatancy properties were assumed to be the same as those of soil.

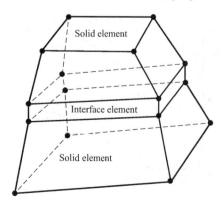
Figure 3.3　3D thin layer element

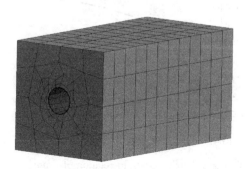
Figure 3.4　Modeling of soil tunnel interfaces in LS-DYNA

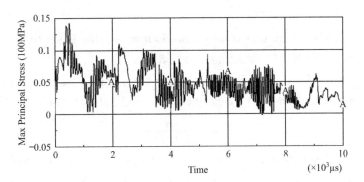

Figure 3.5 Maximum principal stresses in the interface

3.3 Modeling of Unsaturated and Saturated Soils

FHWA soil model is used to model the unsaturated and saturated soil in this study. This model can be applied to different soil types. The model is a modified Drucker-Prager plasticity model and can simulate pre-peak hardening, post-peak strain softening, and pore-water effects (moisture effects).

The modified yield surface is a hyperbola fitted to the Mohr-Coulomb surface. At the crossing of the pressure axis, the modified surface is a smooth surface and it is perpendicular to the pressure axis. The yield surface is given as:

$$F = -P\sin\beta + \sqrt{J_2 K(\theta)^2 + ahyp^2 \sin^2\beta} - c\cos\beta = 0 \qquad (3.3)$$

Where P is pressure, β is model parameter related to internal friction angle, $K(\theta)$ is function of the angle in deviatoric plane, J_2 is the second invariant of the stress deviator, c is amount of cohesion, $ahyp$ is the parameter for determining how close to the standard Mohr-Coulomb yield surface the modified surface is fitted. Figure 3.6 shows comparison of Mohr-Coulomb yield surfaces in the shear stress-pressure space.

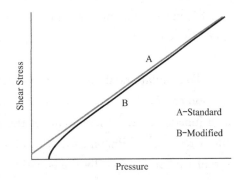

Figure 3.6 Comparison of Mohr-Coulomb yield surfaces in shear stress-Pressure space

To generalize the shape in the deviatoric plane, the standard Mohr-Coulomb $K(\theta)$ function was changed to a function used by Klisinski:

$$K(\theta) = \frac{4(1-e^2)\cos^2\theta + (2e-1)^2}{2(1-e^2)\cos\theta + (2e-1)[4(1-e^2)\cos^2\theta + 5e^2 - 4e]^{1/2}} \qquad (3.4)$$

Where e is material parameter describing the ratio of triaxial extension strength to triaxial compression strength.

Here $\cos(3\theta) = \dfrac{3\sqrt{3}J_3}{2J_2}$, J_2 and J_3 are respectively the second and third invariants of the stress deviator.

If e is set to 1, then a circular cone surface is formed. If e is set to 0.55, then a triangular surface is formed. $K(\theta)$ is defined for $0.5 < e \leqslant 1.0$.

To simulate the effects of moisture and air voids, including excess pore-water pressure, both the elastic and plastic behaviors can be modified. The bulk modulus is:

$$K = \frac{K_i}{1 + K_i D_1 n_{cur}} \tag{3.5}$$

Where D_1 is material constant controlling the stiffness before the air voids are collapsed, $n_{cur} = \text{Max}[0, (w - \varepsilon_v)]$ is the current porosity, $w = n(1-S)$ is volumetric strain corresponding to the volume of air voids, n is porosity of the soil and S is degree of saturation, ε_v is the total volumetric strain.

To simulate the loss of shear strength caused by excess pore-water effects, the model uses a standard soil mechanics technique of reducing the total pressure, P, by the excess pore-water pressure, u, to get an "effective pressure," P': $P' = P - u$

To calculate the pore-water pressure, u, the model uses an equation similar to the equation used for the moisture effects on the bulk modulus:

$$u = \frac{K_{sk}}{1 + K_{sk} D_2 n_{cur}} \varepsilon_v \tag{3.6}$$

Where K_{sk} is bulk modulus for soil without air voids (skeletal bulk modulus) and D_2 is material constant controlling the pore-water pressure before the air voids.

With the D_2 parameter set relatively high compared to K_{sk}, there is no pore pressure until the volumetric strain is greater than the strains associated with the air voids. However, as D_2 is lowered, the pore pressure starts to increase before the air voids are totally collapsed. The K_{sk} parameter affects the slope of the post-void collapse pressure-volumetric strain behavior.

To simulate nonlinear strain hardening behavior, model parameter related to internal friction angel, β, is increased as a function of effective plastic strain, $\varepsilon_{effplastic}$. It is increased as a function of E_t, the amount of the nonlinear strain hardening effects desired, and A_n, the percentage of Phimax where nonlinear behavior begins. The increase in the angle of internal friction is given by the equation:

$$\Delta \beta = E_t \left(1 - \frac{\beta - \beta_{init}}{A_n \beta_{max}}\right) \Delta \varepsilon_{effplastic} \tag{3.7}$$

For input in LS-DYNA, A_n is expressed as a decimal, with values between 0 and 1.0 (0 percent and 100 percent). E_t affects the rate at which nonlinear hardening occurs.

3.4 Evaluation of FHWA Soil Model

3.4.1 Introduction

This section aimed to evaluate the FHWA Soil Model under different loading paths.

The following basic properties of soil were assumed in the evaluation.

Soil properties　　　　　　　　　　　　　　　　　　　　　Table 3.1

Soil type	sand
Peak friction angle	35°
Residual friction angle	30°
Cohesion	6.5 kPa
Void ratio	1.0
γ_{sat}	18 kN/m³
Bulk modulus K (at $\sigma'_m = 100$kPa)	32.4 MPa
Shear modulus G (at $\sigma'_m = 100$kPa)	19.4 MPa
Specific gravity G_s	2.684

The following evaluations are carried out:

The responses of the unsaturated and saturated soil in triaxial compression during monotonic loading, unloading and reloading;

The responses of the unsaturated and saturated soil in isotropic compression, during monotonic loading, unloading and reloading;

A series of simulation in LS-DYNA were conducted to evaluate the following model parameters: M_{cont}, D_1, D_2, K_{sk}, G, A_n, E_t, ξ_0 and G_f. The list of model parameters in LS-DYNA is shown in Table 3.2. Valuable conclusions were found about application of this model.

The list of model parameters　　　　　　　　　　　Table 3.2

Elastic and soil characteristics	K	G	M_c	ρ	γ_{sp}
Plasticity	φ	c	$ahyp$	e	
Pore water effects	D_1	K_{sk}	D_2		
Strain Hardening	A_n	E_t			
Strain Softening	ξ_0	G_f	φ_{res}		
Strength Enhancement Caused by strain-rate effects	γ	n			
Element Deletion	$damlev$	$epsmax$			
Miscellaneous	$nplot$	$rhowat$	$itermax$		

Note: Details of these parameters can be found in Manual for LS-DYNA Soil Material Model 147.

3.4.2 Evaluation

1. Parameter: Moisture content (M_c)

Isotropic compression is a common exercise in soil testing. Isotropic compression tests were conducted here to evaluate the effects of water content. Monotonic loading process, unloading and reloading cases were tested. This model could accurately simulate the process of isotropic compression test and the results accorded well with the soil behavior in lab tests.

Table 3.3 shows the model parameters of saturated soil. All the model parameters were set up to be the same except the water content during the evaluation for the unsaturated soil. From Figure 3.7, at the same volumetric strain, saturated soil sample can afford more pressure than the dry soil. It means that saturated soil sample produces less volumetric strain than dry soil at the same level of pressure. This is quite reasonable since the water is more incompressible than soil skeleton.

Model parameters for isotropic compression test ($M_{cont}=0.36863$, $S=1$) Table 3.3

*MAT_FHWA_SOIL							
mid	ro	nplot	spgrav	rhowat	vn	gammar	intrmx
1	1.84E-6	3	2.684	1E-6	1.1	0	10
k	g	phimax	ahyp	coh	eccen	an	et
0.349	0.208	0.61	1.0E-7	6.5E-6	0.7	0	10
mcont	pwd1	pwksk	pwd2	phires	dint	vdfm	damlev
0.369	56	0.07	286	0.52	1E-5	6E-8	0.8
epsmax							
0.03							

Unit system: mm, kg, ms

Figure 3.8 shows the isotropic compression unloading and reloading cases with different moisture contents. The dry soil sample had more plastic volumetric strain than saturated one. Since water carried part of the total pressure in the saturated soil sample, the effective pressure carried by soil skeleton became smaller and then volumetric strain of the saturated soil was smaller than the dry soil.

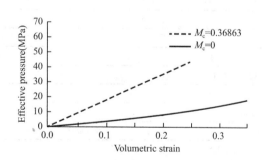

Figure 3.7 Isotropic compression test monotonic loading with different moisture contents

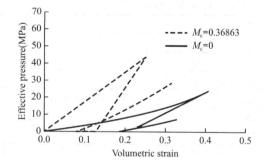

Figure 3.8 Isotropic compression test unloading and reloading with different moisture contents

2. Parameter: D_1

Isotropic compression tests with dry soil were conducted. From the results the soil model can describe the character of soil properties accurately. From the above Equation (3.5) calculating K, the modulus becomes bigger when n_{cur} becomes smaller with compression. So the soil should have hardening effects according to the theory.

Figure 3.9 shows the changes of pressure with volume strain in a monotonic isotropic compression test with dry soil. At the beginning the soil acted elastically since the relationship between pressure and volume strain was linear. When volume strain was larger than 0.15, the pressure increased faster than linear relation, which agreed well with actual soil behavior.

Figure 3.10 shows the changes of pressure with volume strain in an unloading and reloading isotropic compression test with dry soil. There was plastic strain when the load went back to zero. The slope of reloading process was almost the same with the loading process. Figure 3.11 shows that at small strain level the soil deformation was mainly in the elastic stage.

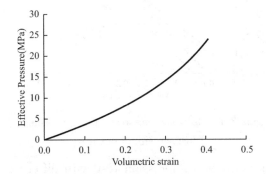

Figure 3.9 isotropic compression test monotonic loading with dry soil

Figure 3.10 Isotropic compression test monotonic unloading and reloading with dry soil

Isotropic compression tests were conducted to testify the effluence of parameter D_1 on the response of soil. The saturate degree of the soil sample was 0.5. The effect of pore water pressure was captured in the analysis. Although the D_1 did not have effects on pore water pressure directly, D_1 had effects on volumetric strain which was related to pore water pressure. From Figure 3.12, it is shown that D_1 increased the bulk modulus. When D_1 increased, at the same level of pressure, the volumetric strain increased, and the water pressure became larger as shown in Figure 3.13.

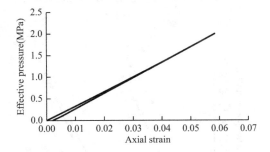
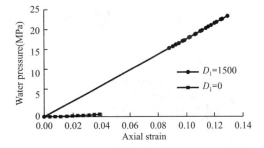

Figure 3.11 Isotropic compression test monotonic unloading at small strain with dry soil

Figure 3.12 Isotropic compression test loading and unloading with $S=0.5$

3. Pore water pressure parameter: K_{sk}

Skeleton bulk modulus was evaluated with different cases: 0.0698GPa, 0.1047GPa and 0.1745GPa. Figure 3.14 shows effective pressure in isotropic compression test with different K_{sk}. The slope of unloading was smaller than the loading process and the reloading slope was the same with the loading case. As K_{sk} increases, the slope of the curve decreases. Since the total pressure applied on the soil samples was the same value, larger K_{sk} produced smaller effective pressure and then generated the larger pore water pressure. This conclusion from simulation matches the theory according to the Equation (3.6).

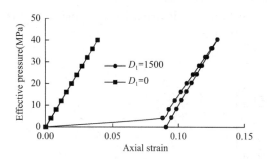

Figure 3.13 Isotropic compression test loading and unloading with $S=0.5$

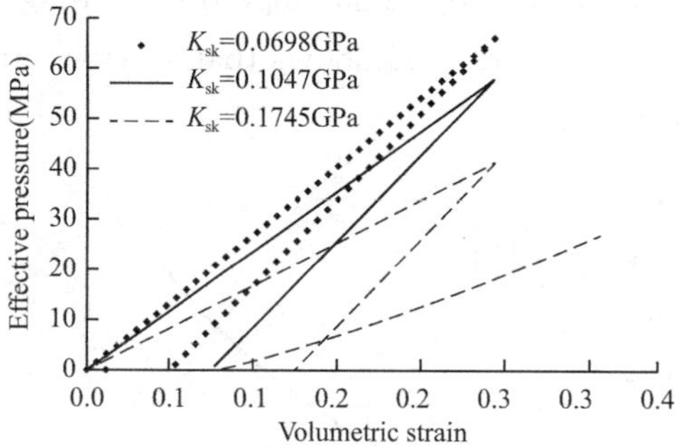

Figure 3.14 Isotropic compression test unloading and reloading with different K_{sk}

Figure 3.15 shows pore water pressure in isotropic compression test with different K_{sk}. The relation between pore water pressure and volumetric strain was almost linear. These results proved the conclusion above: pore water pressure increases with K_{sk}.

4. Pore water pressure parameter: D_2

Different values of D_2 were tried while all the other parameters remained the same values. As shown in Figure 3.16, when D_2 was smaller, in a same process of loading and unloading, the soil sample produced less axial strain. Figure 3.17 shows the changing of pore water pressure with axial strain. A smaller D_2 value produced more water pressure and this trend accorded well with the equation (3.6) in the model. With the D_2 parameter set up to 1500, there is no pore pressure until the volumetric strain is greater than the strains associated with the air voids. However, as D_2 was 500 the pore pressure started to increase earlier than that in the case $D_2=1500$.

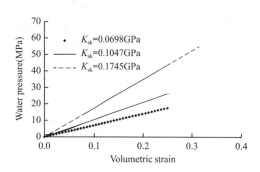
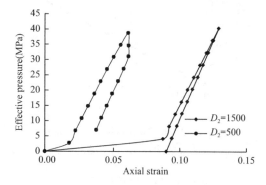

Figure 3.15 Water pressure in isotropic compression test with different K_{sk}

Figure 3.16 Effective pressures in isotropic compression test with different D_2

5. Strain hardening parameter: A_n

It was shown that soil needed more strain to reach the peak shear strength when A_n increased. This meant that soil hardening increased with A_n, as shown in Figure 3.17.

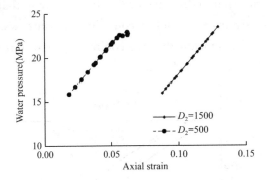

Figure 3.17 Water pressures in isotropic compression test with different D_2

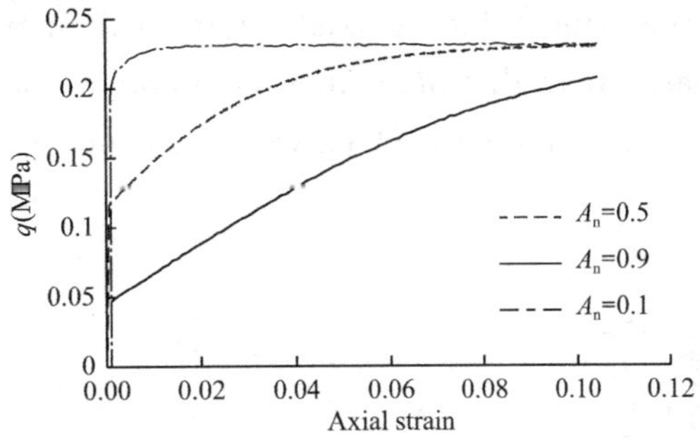

Figure 3.18 Triaxial compression test with dry soil when $A_n = 0.1, 0.5, 0.9$

The definition of q is shown in the equation:

$$q = \frac{\sigma_1 - \sigma_3}{2} \tag{3.8}$$

With A_n was set up to 0.9, q did not reach the peak value even the axial strain was already 0.1.

From Figure 3.19, when A_n was set up to 0.1, the relationship between volume strain and axial strain was close to linear relation. When A_n was set up to 0.9, there were more nonlinear effects on volume strain. When A_n increased, at the same axial strain the volume strain decreased.

6. Strain hardening parameter: E_t

A_n was fixed at 0.5 and different values of E_t were tried to evaluate effect of this parameter. It was shown that soil needed more strain to reach the peak shear strength when E_t decreased. This meant that the soil hardening increased when E_t decreased.

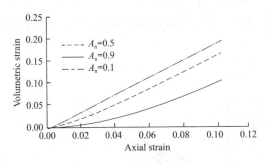
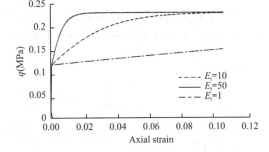

Figure 3.19 Volume strain with axial strain in triaxial compression test when $A_n = 0.1, 0.5, 0.9$

Figure 3.20 Triaxial compression test in LSDYNA with dry soil when $E_t = 1, 10, 50$

From Figure 3.21, the dilatancy of soil under shearing decreased with a decreasing in E_t. This was consistent with the increasing of soil stress with E_t. With an increasing in E_t, the hardening of soil decreased, and the dilatancy of soil also increased.

7. Effects of confining pressure

Confining pressure increased the bulk modulus and shear strength of the soil. Two different confining pressures were conducted to check the effects of confining pressure. Figure 3.22 shows the developing process of volumetric strain with axial strain. It was shown that at the same value of axial strain, volumetric strain with higher confining pressure was smaller than that of smaller confining pressure. This meant that the confining pressure reduced soil dilation, which was consistent with actual soil behavior.

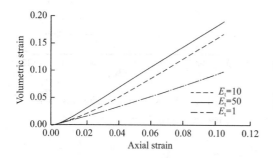

Figure 3.21 Volume strain with axial strain in triaxial compression test when $E_t = 1, 10, 50$

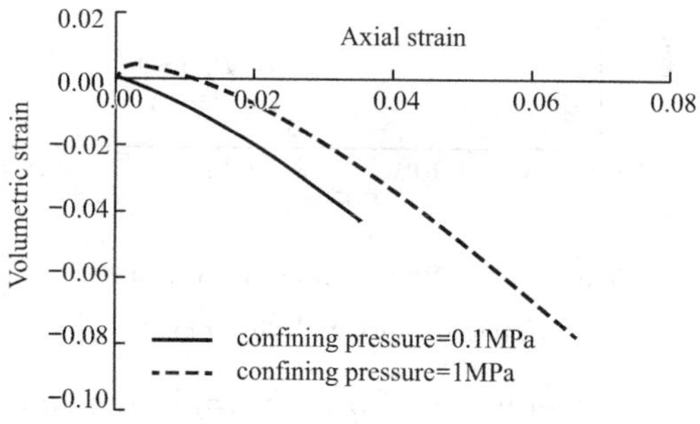

Figure 3.22 Results of trixial compression test with different confining pressure

Figure 3.23 shows the development of deviatoric stress with axial strain. It was shown that at the same value of axial strain, deviatoric stress with confining pressure was much larger than that with smaller confining pressure. This meant that the confining pressure increased the shear strength, which was also consistent with actual soil behavior.

8. Residual pore water pressure upon unloading

Isotropic compression test with cyclic loading was carried out on saturated soil sample to check if the model could catch the process of residual excess pore pressure after unloading. It is from Figure 3.24 that there was plastic volumetric strain after unloading. This residual volumetric strain implied residual pore water pressure according to Equation (3.6). So this model can be used to study the effects of residual pore water pressure on the tunnel response after internal blast loading.

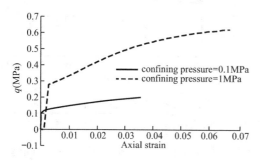
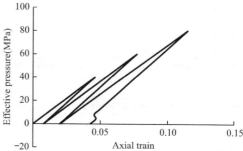

Figure 3.23 Results of trixial compression test with different confining pressure

Figure 3.24 Results of cycle loading and reloading isotropic compression test

3.5 Conclusions

An isotropic elastic-plastic material model in LS-DYNA was used to simulate the cast-iron tunnel lining in the numerical analysis. In this model yield stress versus plastic strain curves can be defined for compression and tension. Thus this model can simulate the cast-iron whose compressive strength is much larger than tensile strength. Thin-layer elements were used to simulate the interface between soil and tunnel. The strength of soil-lining interface was assumed to be 2/3 of soil strength, while its stiffness and dilatancy properties were assumed to be the same as those of soil.

*MAT_FHWA_SOIL material model is designed to specifically to predict the dynamic performance of the foundation soil with road safety structures. This model can be applied to different soil types. The model is a modified Drucker-Prager plasticity model and can simulate pre-peak hardening, post-peak strain softening (damage), strain-rate effects (strength enhancement), pore-water effects (moisture effects), and erosion capability. This soil model can simulate the effect of pore water pressure, volumetric compression, and shear strain hardening, which is relevant to this study. Particularly, residual pore water pressure and soil liquefaction after large blast loading may be modeled using this model.

However, the model can only trace pore water pressure when volumetric strain is compression. When the soil sample has dilation, the pore water pressure in this model is set up to zero.

Chapter 4 Response of Circular Tunnels Subjected to Medium Internal Blast Loading

4.1 Base-Case Finite Element Model

As shown in the Figure 4.1, the Finite Element model includes soil, soil-tunnel interface, tunnel, air and ambient layer. Due to symmetry, 1/4 model was simulated to save computer resources. The prototype model is based on single-track subway tunnels in New York City.

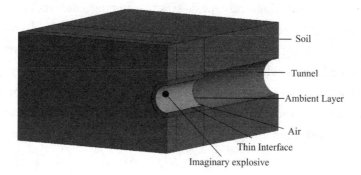

Figure 4.1 Base-Case Finite Element Model

The diameter of the tunnel was assumed to be 5m and the tunnel was buried 7.5m blow the ground surface. The thickness of saturated soil layer was assumed to be 15m, the base of which was stiff bed rock and fixed in the Finite Element model.

Due to symmetry, the length of the tunnel in the model was 30m. The width of the model was 25m. The lining thickness was assumed to be 6cm based on the parameters of cast-iron subway tunnels in New York City. The Finite Element model was fixed at the base and symmetric boundaries were applied to the symmetric planes respectively.

With a scale factor of 5, the length of tunnel was 6m as shown in Figure 4.2. The tunnel was modeled into three sections with different element sizes along the longitudinal direction. In the first 50cm section the element size was 1cm, in the next 150cm section, the element size was 3cm, and in the last section the element size was 8cm. These Finite Element parameters have been determined by trial analyses. With smaller element sizes, the responses of the tunnel were almost identical.

The density of cast iron tunnel is $7.89g/cm^3$. The Young's modulus of cast iron tunnel is 100GPa. In the tensile direction, the tunnel is in the elastic domain before the strain reaches 0.001. The failure stress is 150MPa when the tensile strain equals to 0.004. In the compressive direction the yield stress of tunnel is 600MPa. The constitutive model and stress-strain curves in tension and compression have been discussed in Chapter 3.

The air domain and ambient layer was modeled as shown in Figure 4.3. The length in the longitudinal direction was 2m which was enough to transfer the blast wave to tunnel, as discussed in Chapter 2. Analysis consisting of air domain with 6m length was also carried out for comparison. The results showed that the air domain as shown in Figure 4.3 was sufficient. The thickness of the air domain was 12cm and the thickness of ambient layer was 0.5cm. Here the air domain and ambient layer were modeled with the same material model parameters in Chapter 2. The air density scale factor was 1.2 (air density $0.001548g/cm^3$).

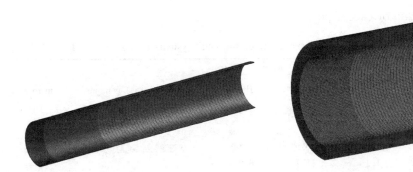

Figure 4.2 Finite Element Model of tunnel in LS-DYNA

Figure 4.3 Finite Element Model of air and ambient layer in LS-DYNA

Figure 4.4 shows the influence of element size on peak pressure and specific impulse in free air blast. At the same scaled distance range, results using element size of 0.4cm accord best with CONWEP solution. However due to the limit of existing computer resource, element size of 1.0cm was adopted in the simulation. Although the peak incident pressure is slightly smaller than but the specific impulse is very close to the CONWEP solutions. Note that the range of scaled distances investigated in this study is larger than $0.42m/kg^{1/3}$.

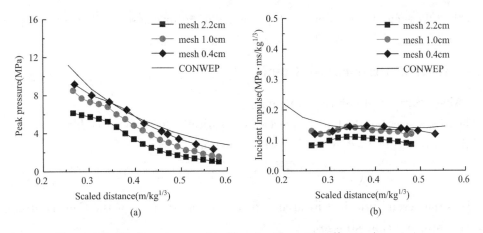

Figure 4.4 Peak pressure and incident impulse with different element sizes

The interface was meshed similarly as the tunnel. The thickness of the interface is 1.2cm in the LS-DYNA model as shown in Figure 4.5. The shear modulus and bulk modulus of the thin layer were set up as two thirds of those parameters of soil. The thin layer and soil were modeled with FHWA soil model. For the saturated soil case, the density of the soil was 2.35g/cm³. Peak shear friction angle was assumed to be 38° while the cohesion was assumed to be 6.2kPa. Table 4.1 shows the parameters of the saturated soil, which are basically identical with the default FHWA soil model parameters, except for the parameter K_{sk} that controls the buildup of excess pore pressure.

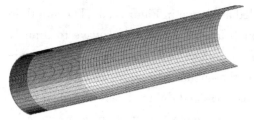

Figure 4.5 Finite element model of thin layer interface.

Material model parameters of the saturated soil (unit system: cm, g, μs) Table 4.1

*MAT_FHWA_SOIL							
mid	ro	nplot	spgrav	rhowat	vn	gammar	intrmx
8	2.35	3	2.79	1E-6	1.1	0	10
k	g	phimax	ahyp	coh	eccen	an	et
0.00465	0.00186	1.1	1E-7	6.2E-8	0.7	0	10
mcont	pwd1	pwksk	pwd2	phires	dint	vdfm	damlev
0.034	0	0.00175	0	0.0	5E-5	1E-9	0.8
epsmax							
0.03							

4.2 Simulation Results with 200kg TNT

In the following, the results are all presented in prototype scale. This section presents the simulation results of 200kg TNT case with saturated soil.

4.2.1 General Response of Tunnel

1. Mises Stress and failure in the lining

The Mises stress in the lining increased dramatically due to internal blast loading. The Von Mises stress is defined in the following equation:

$$\sigma_v = \sqrt{3J_2} = \sqrt{\frac{(\sigma_1-\sigma_2)^2 + (\sigma_2-\sigma_3)^2 + (\sigma_3-\sigma_1)^2}{2}} \quad (4.1)$$

Here J_2 is the second principal invariants of the deviatoric part of the Cauchy stress. σ_1, σ_2, σ_3 are three principle stresses.

Figure 4.6 shows the time history of Von Mises stress in the lining at different locations. When the tunnel reached the failure stress of 220MPa, the lining element failed and was deleted from further analysis. Figure 4.7 shows deleted elements distribution in the tunnel. The failure mainly occurred in the first 2.5m section in prototype scale. After 2.7ms the failure extended to the farther distance than 2.5m.

Failure began to appear at 1.125ms. The blast wave reached the tunnel at 0.25ms and the reflected blast pressure in the air element next to tunnel showed the first peak value at

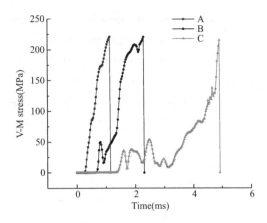

Figure 4.6 Time history of $V\text{-}M$ stress in the failed lining elements (location of points A B C is shown in figure 4.7a)

0.425ms. The reflected blast wave reduced to ambient pressure when the failure appeared, as shown in Figure 4.8. It appears that the tunnel was more affected by the specific impulse of impact rather than by the peak pressure.

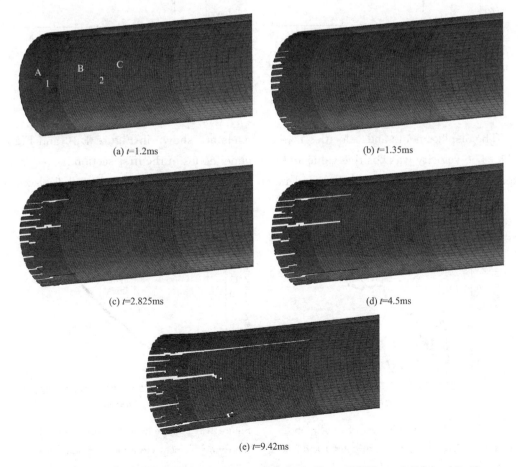

(a) t=1.2ms

(b) t=1.35ms

(c) t=2.825ms

(d) t=4.5ms

(e) t=9.42ms

Figure 4.7 Failure of tunnel elements at t=1.2ms, 1.35ms, 2.825ms, 4.5ms and 9.42ms

2. Displacement, velocity and acceleration in the lining

The assumed termination time of the analysis was set up to 15ms. However in this case the analysis terminated at 9.42ms due to extensive failure in the tunnel and negative pore water pressure in the soil.

The acceleration in the tunnel is shown in Figure 4.9. The tunnel experienced a large acceleration under blast loading. The maximum acceleration of point 1 in the expanding direction was 67132m/s^2 and the maximum acceleration in the shrinking direction was 55757m/s^2.

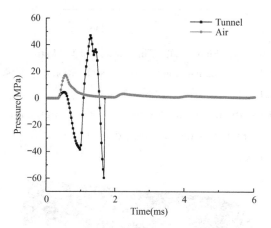

Figure 4.8 Time history of pressure in the adjacent air element and lining element (lining element is located at point a in figure 4.7a)

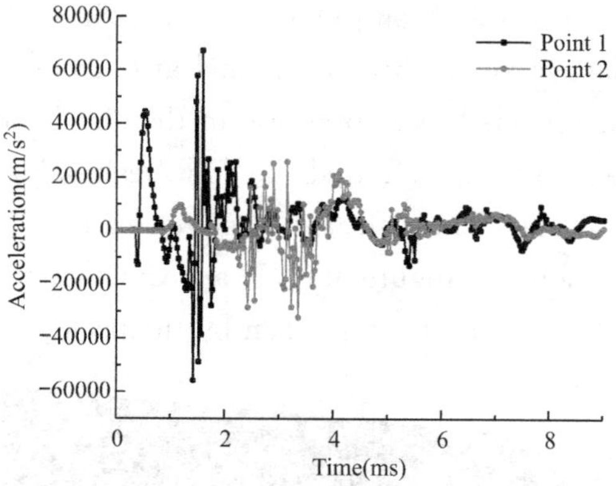

Figure 4.9 Time history of acceleration in the tunnel element (locations of point 1 and 2 was shown in the Figure 4.7a)

The displacements and velocities in the tunnels are shown in Figure 4.10 and Figure 4.11. The velocity was positive value and the tunnel nodes in the first section moved in the expanding direction. Although the acceleration became negative, the velocity and displacement

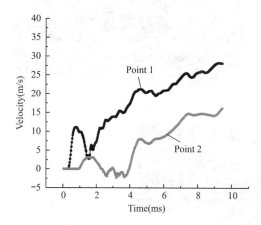
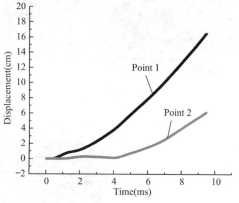

Figure 4.10 Time history of velocity in the tunnel element (locations of point 1 and 2 was shown in the Figure 4.7a)

Figure 4.11 Time history of displacement in he tunnel element (locations of point 1 and 2 was shown in the Figure 4.7a)

were both of positive values, which meant that the tunnel mainly displaced in the expanding direction. The maximum velocity of point 1 was 28m/s and the maximum velocity of point 2 was 16m/s. The maximum displacement of the point 1 was 16.45cm and the maximum displacement of point 2 was 6.09cm. Both of them were in the expanding direction.

The acceleration, velocity and displacement all decreased significantly in the lining where it was farther away from the source of blast loading. The responses were very small in the lining that was more than 10m away from the blast source.

3. Plastic strain in the lining

The effective plastic strain distributions at different moments after blast loading are shown in Figure 4.12. The definition of effective plastic strain is shown in the following equation:

$$\varepsilon_{eff}^p = \int_0^t d\varepsilon_{eff}^p \qquad (4.2)$$

where $d\varepsilon_{eff}^p = \sqrt{\frac{2}{3} d\varepsilon_{ij}^p d\varepsilon_{ij}^p}$.

The trend was similar to that of Mises stress. In Figure 4.12, the deformed shape of the tunnel is also illustrated (enlarged to 20 times). The deformation of the tunnel lining can be better witnessed in this figure. Figure 4.13 shows the time histories of plastic strain at different locations in the lining. Where there was no failure, the plastic strain remained at a constant value certain time after blast loading, indicating unloading at the particular location.

4. Impact loading on tunnel lining

The peak pressure and specific impulse on the lining that did not fail are shown in Figure 4.14. The peak pressures were smaller than the reflected one on a rigid plane while the impulses were larger. Comparing with the results in Chapter 2, the peak pressures were smaller while the impulses were larger than those in Chapter 2. This difference was caused by the deformation of tunnel in this chapter and the multiple reflections of air pressure herein.

The pressure time histories at three locations are shown in Figure 4.15. The angels of incidence were 0°, 26.6° and 51.3°, respectively. For the 0° case, the second peak was only 10% of the first peak. When the incident angle became 51.3°, the second peak was almost the same as the first peak. However the response of tunnel was very small when the incident angle was large.

4.2.2 Response in Soil-Tunnel Interface

In the analysis the blast wave reached the interface between soil and tunnel at 0.575ms after the explosion. The blast wave propagated along the longitudinal direction and the magnitude of the V-M stress in the thin layer increased with time. However the influence of impact wave was mainly inside the domain not far away from the explosive (about 5.5m away from the explosive in the longitudinal direction). The thin layer deformed dramatically in response to the blast wave loading in this domain as shown in the Figure 4.16. The thin layer vibrated with the tunnel with a smaller magnitude. V-M stresses distribution in the thin layer during the analysis is shown in Figure 4.16. Separation was observed in some part of the tunnel but it was not extensive.

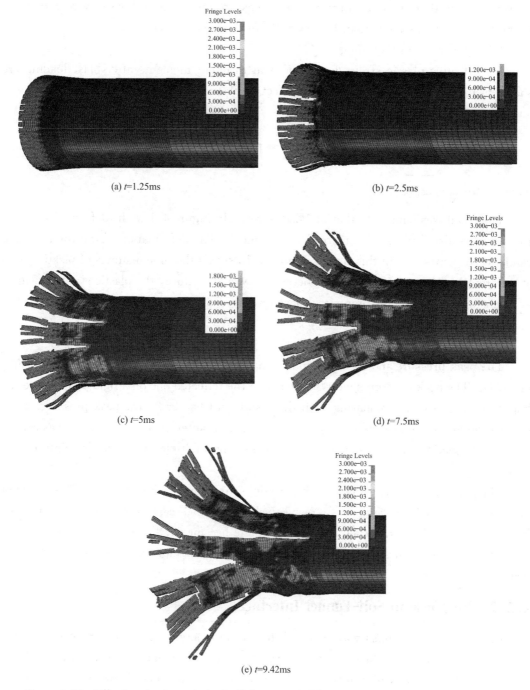

Figure 4.12 Effective plastic strain in the lining at $t=1.25$ms, 2.5ms, 5ms, 7.5ms and 9.42ms

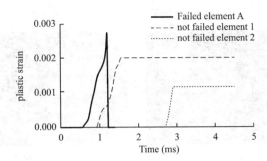

Figure 4.13 Time history of plastic strain in the tunnel elements (locations of element A, 1 and 2 are shown in Figure 4.7a)

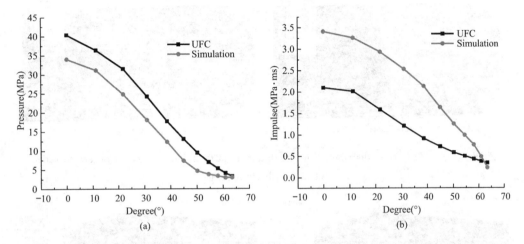

Figure 4.14 Peak pressures and impulses at different incident angles in the air elements next to tunnel lining

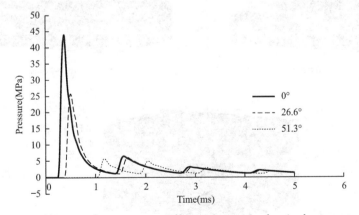

Figure 4.15 Time history of pressures at different degrees in the air elements next to tunnel

4.2.3 Response in Soil

Figure 4.17 shows the effective strain distribution inside the soil. The effective strain is defined as:

$$\varepsilon_{\text{eff}} = \sqrt{\frac{2}{3} \mathrm{d}\varepsilon_{ij}\, \mathrm{d}\varepsilon_{ij}} \tag{4.3}$$

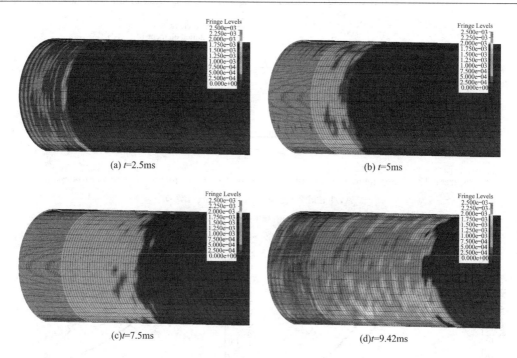

Figure 4.16 V-M stress distribution in the thin layer at $t=2.5$ms, 5ms, 7.5ms and 9.42ms. (Units: Mbar=1011Pa)

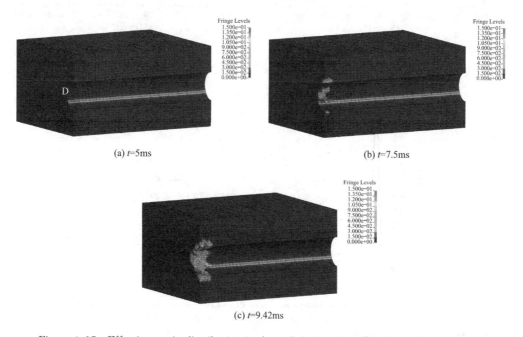

Figure 4.17 Effective strain distribution in the soil during the analysis at $t=5$ms, 7.5ms and 9.42ms (deformation enlarged 2 times).

The blast wave reached soil at 1.275ms after the explosion. The blast wave propagated in both the longitudinal direction and axial direction. The main influence of impact load occurred in the soil not far away from the explosive. The soil expanded first and then vibrated. The soil deformation was enlarged with a scale factor of 2 for better clarity. The maximum effective strain in the soil close to the blast loading was 23%. In addition, although not clearly demonstrated herein, some soil elements were penetrated by the fractured tunnel lining.

The pressure output from LS-DYNA was effective stress in the FHWA soil material model. If the effective stress decreases to zero, it means that liquefaction occurred as shown in Figure 4.18. Figure 4.19 (a) ~ (c) shows the liquefaction process inside the tunnel. Liquefaction was concentrated in the soil near the explosive. As far as 2.5m away in the longitudinal direction, no liquefaction occurred. The top soil had more liquefaction than the bottom soil due to effect of gravity load.

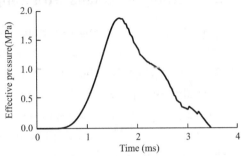

Figure 4.18 Time history of effective pressure in the liquefied soil element. (Location D as shown in Figure 4.17a)

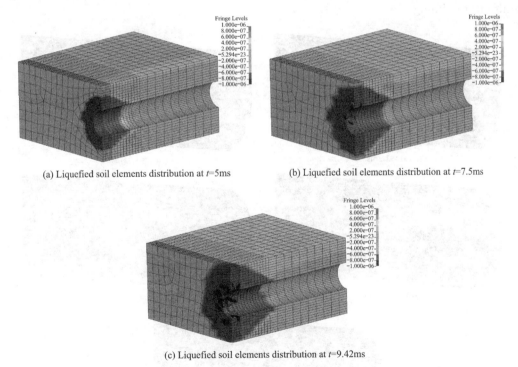

(a) Liquefied soil elements distribution at t=5ms

(b) Liquefied soil elements distribution at t=7.5ms

(c) Liquefied soil elements distribution at t=9.42ms

Figure 4.19 Liquefied soil elements distribution

The soil was modeled by the FHWA Soil Model in this study. The constitutive model assumes associated flow rule and the model soil dilates considerably when it starts yielding. This unrealistic dilation would result in reduction of pore water pressure and increase of effective stress. Due to this defect in the constitutive model, a soil element liquefied after the end of blast loading. However, the liquefaction disappeared at the same location when the soil element started yielding and dilating due to small effective stress. The distribution of soil liquefaction in Figure 4.19 (a) ~ (c) was therefore under-estimated by the model.

4.2.4 Analysis Results with Simplified Blast Load

As comparison, an analysis employing the reflected pressure on a rigid surface was also carried out. The blast pressure was applied directly on the lining surface using LOAD_BLAST in LS-DYNA. The Finite Element model of the tunnel-soil system was exactly the same as that in Figure 4.1, but without the air elements.

Figure 4.20 shows the responses of the lining. It can be seen that the lining damage was not as severe. The length of ruptured lining was only half of the previous one. The peak pressure and specific impulse at a point close to the point of explosion (Point A in Figure 4.7a) were extracted from the LBE analysis with air, which were 34MPa and 2.72MPa · ms/kg$^{1/3}$, respectively. The corresponding ones from UFC 3-340-02 (2008), which were also the applied ones by LOAD_BLAST, were 40.4MPa and 1.76MPa · ms/kg$^{1/3}$. The specific impulse with LBE method was much higher, which resulted in the more severe failure in the lining.

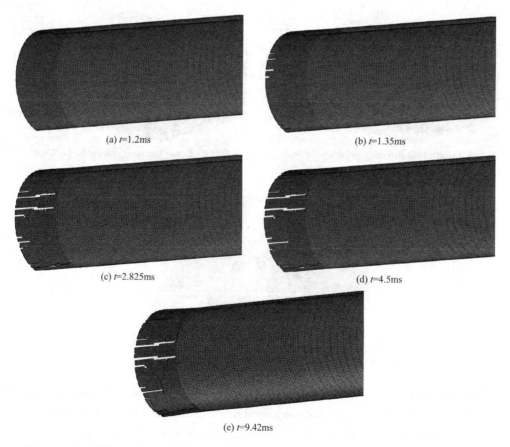

Figure 4.20 Failure of tunnel elements at $t=1.2$ms, 1.35ms, 2.825ms, 4.5ms and 9.42ms

4.2.5 Equivalent Triangle Impact Load

An equivalent triangle impact load was assumed and applied to the tunnel lining directly. As shown in the Figure 4.21, the time histories of impact loads on different domains of tunnel were calculated according to the pressure and specific impulse results generated in ALE analysis.

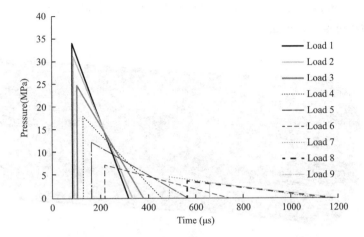

Figure 4.21 Triangle impact load applied on the tunnel equivalent to 200kg TNT

Nine domains of the tunnel as shown in Figure 4.22 were subjected to triangle impact load respectively. The pressure and specific impulse on the other part of the tunnel was very small and had negligible effects on the response of the tunnel. They were therefore neglected in the analysis. The simulation results are shown in Table 4.2.

Figure 4.22 Tunnel to be applied the triangle impact load

results of simulation with triangle impact load and Load _ Blast method   Table 4.2

Simulation results		LBE	Triangle Load	LB Method
Failure area (m²)		16.84	12.89	9.07
Liquefaction volume (m³)		12.13	23.58	13.26
Max shear strain in the tunnel		0.00299963	0.0029996	0.00299987
Max effective shear strain in soil		23%	35%	30%
Plastic strain in the tunnel	1	0.0027	0.0028	0.0026
	2	0.0020	0.0019	0.0018

It can be seen that the triangle impact load resulted in smaller failure in the lining. The failure mostly occurred in the domain near the explosive. The area of ruptured lining was smaller than the one with LBE method. Load _ Blast method without modeling air domain showed the similar trend as the triangle load method, but the ruptured area was even smaller. However the method with assumed triangle impact load could be acceptable, as the tunnel response was not significantly different from that by LBE approach with air domain. When the computer resources are limited, this method can be adopted to simulate effect of internal blast loading.

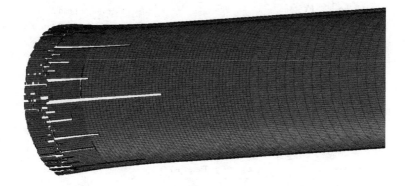

Figure 4.23 Failure in the tunnel at $t=9.42$ms in the 200kg TNT case with impact load method

Plastic strain in the lining at $t=1.25$ms, 2.5ms, 5ms, 7.5ms and 9.42ms with two different loading method　　　　　　　　　　　　　　　　　Table 4.3

Triangle impact load	LBE with air

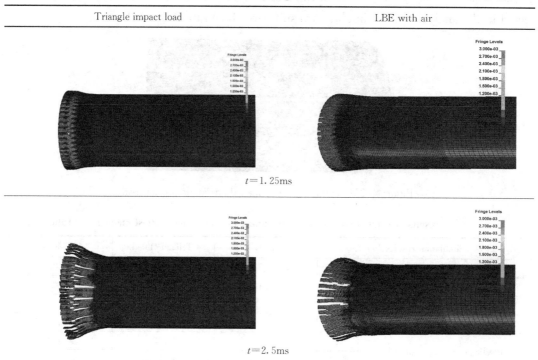

续表

| Triangle impact load | LBE with air |

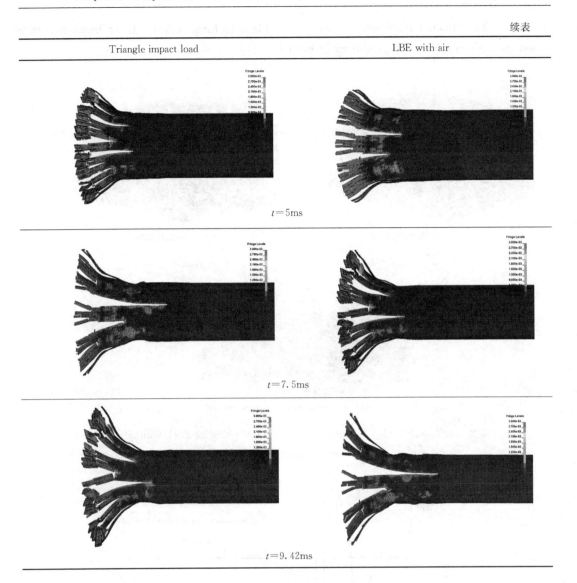

$t=5$ms

$t=7.5$ms

$t=9.42$ms

4.3 Simulation Results with 100kg TNT

This section presents the simulation results with 100kg TNT and saturated soil.

4.3.1 Response of Tunnel

1. Mises Stress and failure in the lining

With 100kg TNT, the lining failure was not as severe as the case with 200kg TNT. The fracture area was about 0.19m² which was much smaller than the 200kg case (13.05m²) in prototype model. The fracture concentrated at the top and bottom of the tunnel. Failure began to appear at 1.5ms. The blast wave reached the tunnel at 0.35ms and the reflected blast pressure in the air element next to tunnel showed the first peak value at

0.55ms. The reflected blast wave close to the blast loading reduced to ambient pressure when the failure appeared, as shown in Figure 4.24.

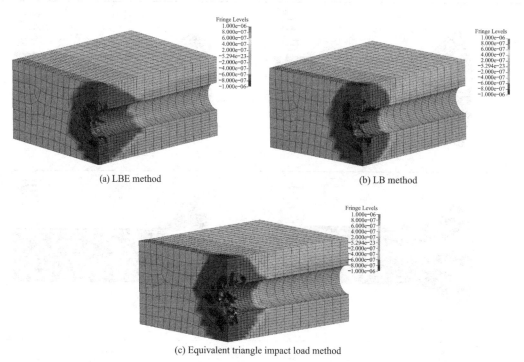

Figure 4.24 Soil Liquefaction distribution at $t=9.42$ms of 200kg TNT with different methods
(The blue parts are liquefied soil)

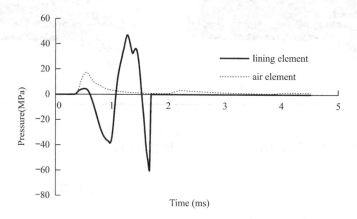

Figure 4.25 Time history of pressure in the adjacent air element and lining element
in 100kg TNT case (lining element is located at point a in figure 4.7a)

2. Displacement, velocity and acceleration in the lining

The acceleration in the tunnel is shown in Figure 4.26. The tunnel experienced a large acceleration under blast loading. In the 200kg TNT case the tunnel failed extensively and the analysis stopped at 9.42ms. In the 100kg TNT case, the analysis continued until the termination time of 15ms. The maximum acceleration in the expanding direction was

80688m/s² and the maximum acceleration in the shrinking direction was -88284m/s². They were larger than the previous case due to the less-extensive failure in the lining.

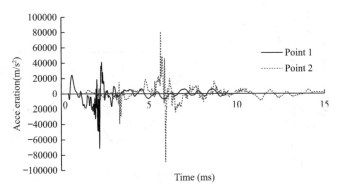

Figure 4.26　Time history of acceleration in the tunnel element (locations of point 1 and 2 was shown in the Figure 4.7a)

The displacements and velocities in the tunnels are shown in Figure 4.26 and Figure 4.27. The peaks of acceleration and velocity of point 2 was larger than point 1. That accorded well with the failure distribution. The fracture occurred close to point 2.

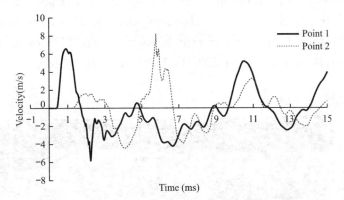

Figure 4.27　Time history of velocity in the tunnel element (locations of point 1 and 2 was shown in the Figure 4.7a)

The maximum displacement of point 1 was 0.51cm in expanding direction and 0.92cm in the shrinking direction. The maximum displacement was 0.19cm in expanding direction and 0.27cm in the shrinking direction for point 2. The lining near the explosive vibrated significantly but did not fracture.

It appears that the lining fracture was caused by the phase lag of vibration, as can be seen in Figure 4.26 and Figure 4.28. The directions and magnitudes of lining displacements and accelerations at different locations were different, resulting in large stress and failure at certain location. This phase lag was mostly initiated by the different moments of blast loading on the lining.

3. Plastic strain in the lining

The plastic strain distributions at different moments after blast loading are shown in Figure 4.29. Similarly the deformed shape of the tunnel is also illustrated (enlarged to 20 times). The deformation of the tunnel lining can be better witnessed in this figure.

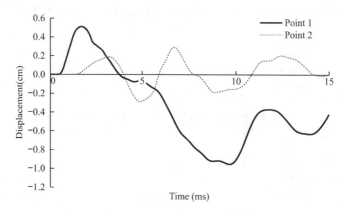

Figure 4.28 Time history of displacement in the tunnel element
(locations of point 1 and 2 was shown in the Figure 4.7a)

4.3.2 Impact Loading on Tunnel Lining

The peak pressure and impulse in the air next to tunnel are shown in Figure 4.30. The peak pressures of simulation at different incident angels were smaller than ones from the UFC 3-340-02 (2008) and this trend was similar as the 200kg TNT case. Then LBE method included more reflection impulses than the UFC 3-340-02 (2008).

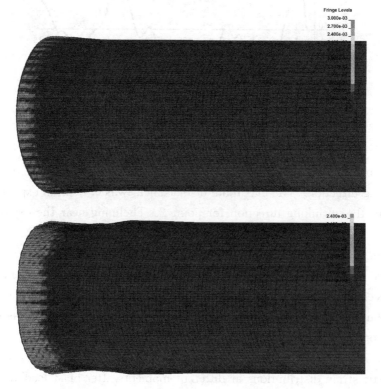

Figure 4.29 Plastic strain in the lining in 100kg TNT case at $t=1.35$ms, 4.5ms, 10ms, 15ms (one)

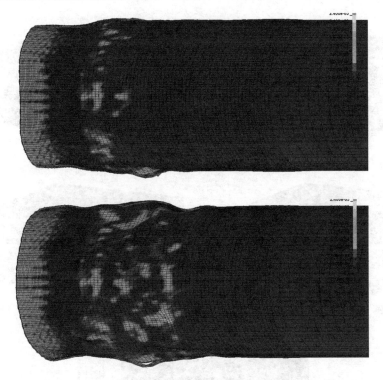

Figure 4.29　Plastic strain in the lining in 100kg TNT case at $t=1.35\text{ms}$, 4.5ms, 10ms, 15ms (two)

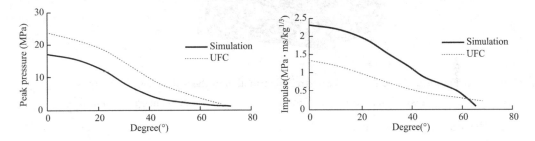

Figure 4.30　Peak pressure and impulse in the air next to tunnel

4.3.3　Response in Soil

1. Effective shear strain inside the soil

Figure 4.31 shows the effective plastic strain distribution inside the soil. The blast wave propagated in both the longitudinal direction and axial direction.

The whole analysis time was 15ms and then the blast wave reached ground surface and the fixed model base. The main influence of impact load occurred in the soil no farther than 7m away from the blast loading. The soil expanded first and then vibrated. The displacement was enlarged with a scale factor of 2 to view the deformation of soil clearly. The region of large shear deformation was larger in this case than in the case with 200kg TNT. With limited damage in the lining, the lining vibrated more significantly and propagated more energy to the surrounding soil.

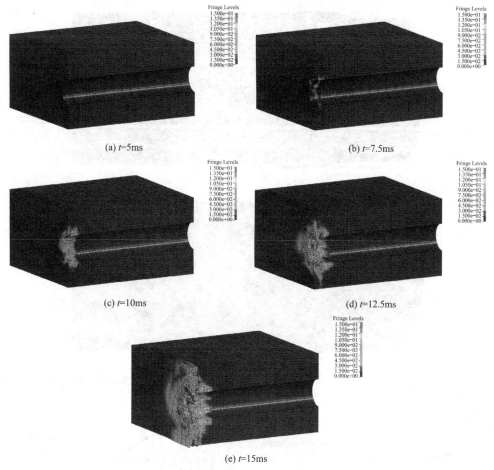

Figure 4.31　Effective shear strain distribution in the soil during the analysis at $t=$5ms, 7.5ms, 10ms, 12.5ms and 15ms (Deformation enlarged 2 times)

2. Residual pore water pressure

Soil liquefaction distribution at $t=$15ms is shown in Figure 4.32. Figure 4.32 (a) ～ (e) shows the liquefaction process inside the soil. The maximum shear strain inside the soil was 34%. The soil began to liquefy at about 4ms and the liquefaction extended to the top

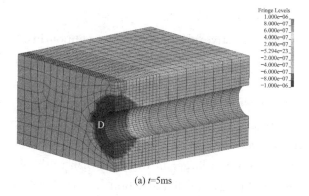

(a) $t=$5ms

Figure 4.32　Liquefied soil elements distribution in 100kg TNT case (one)

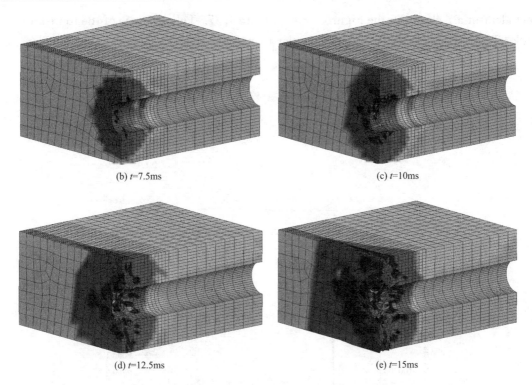

Figure 4.32 Liquefied soil elements distribution in 100kg TNT case (two)

and bottom of the soil. The time history of effective in soil element D was shown in Figure 4.33. The soil liquefaction mainly occurred within the 7m far away from the explosive in the longitudinal direction.

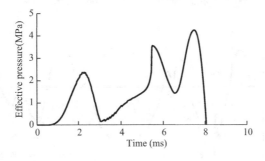

Figure 4.33 Time history of effective pressure in the liquefied soil element. (Location of D is shown in Figure 4.32a)

4.4 Simulation Results with 50kg TNT

4.4.1 Response of Tunnel

The tunnel had no fracture but some tunnel elements near the explosion failed, as shown in Figure 4.34. The time histories of acceleration, velocity and displacement of tunnel element are shown in the Figure 4.35 ~Figure 4.37. The response of the tunnel due to 50kg TNT equivalent explosive was smaller than that in the 100kg and 200kg TNT cases.

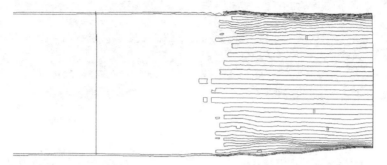

Figure 4.34   Failure extension length in the tunnel of 50kg case at $t=$ 15ms

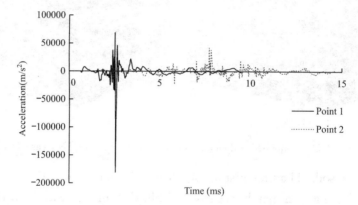

Figure 4.35   Time history of acceleration in the tunnel element (locations of point 1 and 2 were shown in the Figure 4.7a)

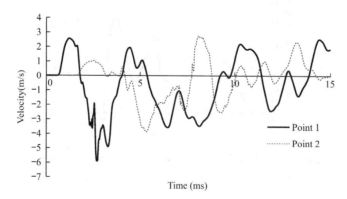

Figure 4.36   Time history of velocity in the tunnel element (Locations of point 1 and 2 were shown in the Figure 4.7a)

4.4.2   Response of Soil

Figure 4.38 shows the effective shear strain distribution of the saturated soil at different moments. The effective shear strain was a little smaller than the 100kg case while the distribution was similar. Figure 4.39 showed the liquefied soil elements in the analysis. The extension of soil liquefaction was smaller than the 100kg case, but larger than the 200kg case. The maximum shear strain was 37%. It appears that most energy from the blast loading propagated into the soil, while in the 200kg case, only a small portion of it propagates into the soil due to extensive failure of tunnel lining.

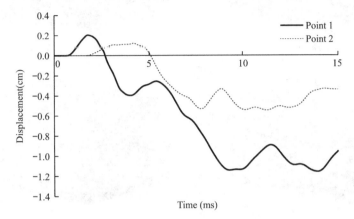

Figure 4.37 Time history of displacement in the tunnel element
(Locations of point 1 and 2 were shown in the Figure 4.7a)

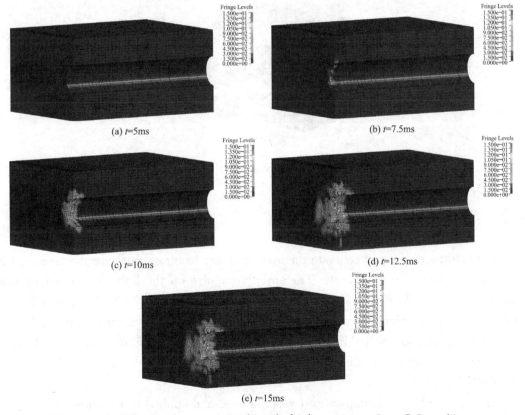

Figure 4.38 Effective shear strain in the soil of 50kg case at $t=$5ms, 7.5ms, 10ms, 12.5ms and 15ms (deformation enlarged 2 times)

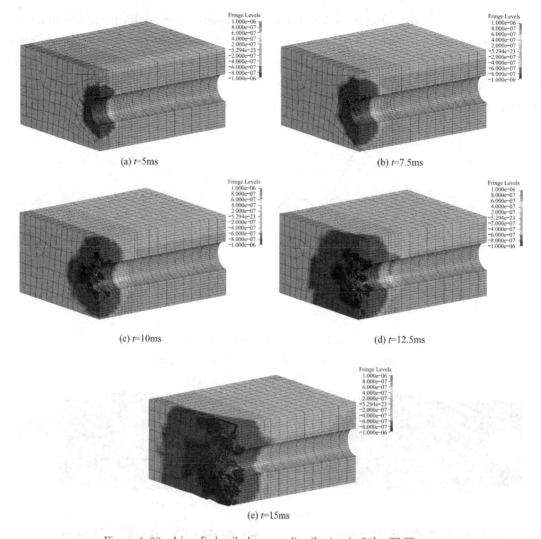

Figure 4.39 Liquefied soil elements distribution in 50kg TNT case

4.5 Conclusions

The LBE method could generate the blast pressure loading accurately and transfer the blast wave to the tunnel properly. The specific impulse on the lining generated by this scheme was larger than the corresponding reflected results from UFC 3-340-02 (2008). The tunnel had confining effects and generated larger impact loading than explosion in free air. The LBE method resulted in more realistic and more severe failure in the tunnel lining when compared with the direct application of reflected blast pressure on the lining surface from CONWEP.

An equivalent triangle impact load was assumed and applied to the tunnel directly to save computer resources. The time histories of impact loads on different domains of tunnel were calculated according to the pressure and specific impulse results generated in the LS-

DYNA. It appears that the simplified method with triangle impact load could be acceptable and can be adopted to simulate internal blast loading inside a circular tunnel.

The failure conditions of tunnel lining were different due to different amounts of explosive. With relatively large amount of explosive, severe rupture firstly appeared in the domain close to the explosive, and then propagated to farther distance due to lining vibration. The failure was governed by the tensile strength of the material. When the blast loading were reduced, only a little fractures might occur. The fracture concentrated in the section 7m away from the explosive. The lining fracture was caused by the phase lag of vibration. The directions and magnitudes of lining displacements and accelerations at different locations were different, resulting in large stress and failure at certain location. This phase lag was mostly initiated by the different moments of blast loading on the lining. Overall the damage and failure of the tunnel lining was progressive in nature. The damage and failure occurred mainly during the lining vibration when the main event of blast loading was over.

The main influence of blast loading occurred in the soil not far away from the explosive. The soil expanded first and then vibrated. Some soil elements were penetrated by the fractured tunnel lining. Soil liquefaction occurred with blast loading from 50~200kg TNT equivalent. Soil had progressive failure due to both vibration of the tunnel and blast loading.

With large amount of explosive and extensive lining failure, only a small portion of blast energy propagated into the soil, and the extension of soil liquefaction was actually smaller. In contrast the soil liquefaction might be more extensive when the lining failure was less severe. The lining vibrated more significantly and propagated more energy to the surrounding soil.

However the excess water pressure generated in the FHWA soil material model was not 100% accurate after the soil liquefied and underwent large shear deformation. Unrealistic dilation of model soil with large shear deformation led to unrealistic increase of effective stress and shear strength. Tunnel might experience more failure due to the soil liquefaction, but it could not be reproduced in this study. More work is needed to improve the material model and the numerical scheme.

Chapter 5   Parametric Study

5.1 Lining Strength and Stiffness

Under internal blast loading, the effects of loading rate and temperature increase may alter the strength of tunnel lining. A series of simulations were carried out to investigate the influence of lining strength and stiffness on the response. The lining strength and stiffness was changed to 110%, 90%, 80% and 70% of the original value. All the other parameters remained the same in the 100kg TNT equivalent cases.

The damage of tunnel lining obviously changed with different tunnel strengths. Figure 5.1 shows the damages of the tunnel lining due to blast loading of 100kg TNT. When the tunnel strength was increased to 110% of the basic case, the tunnel had no failure and the deformation was smaller than the base case.

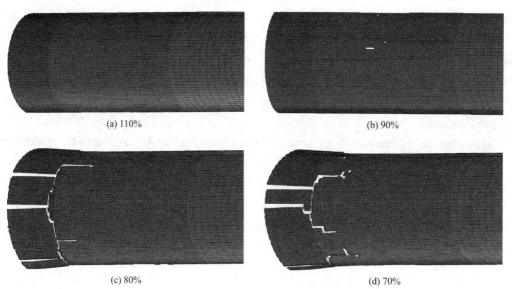

(a) 110%   (b) 90%

(c) 80%   (d) 70%

Figure 5.1   Failures in the tunnel of 100kg TNT with different tunnel strength and stiffness (110%, 90%, 80% and 70%)

When the strength scale factor was assumed to be 90%, the failure mode was basically the same as the one with the original strength. But the failure area and maximum plastic strain were smaller than the base case. However when the strength scale factor was even smaller, rupture firstly appeared in the section close to the explosive, and then propagated to farther distance due to lining vibration. The failure areas and maximum plastic shear strains of tunnel at $t=15$ms are shown in Figure 5.2 and Figure 5.3, respectively.

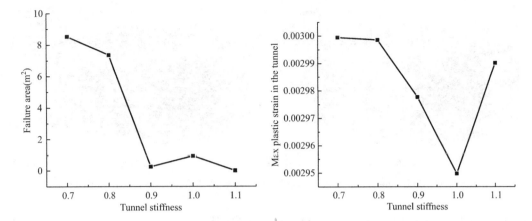

Figure 5.2　Failure areas in the tunnel with different tunnel stiffness

Figure 5.3　Maximum plastic strains in the tunnel with different tunnel stiffness

Figure 5.4 and Figure 5.5 show the maximum effective shear strain and liquefied soil volume due to different tunnel strengths. The soil had less liquefaction and smaller deformation when the tunnel strength decreased. Figure 5.6 shows the liquefaction distribution in the soil at $t=15$ms (The blue parts were the liquefied soil).

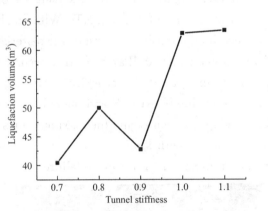
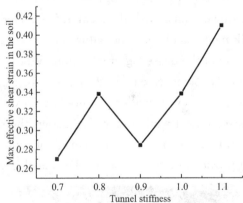

Figure 5.4　Liquefied soil volumes in the tunnel with different tunnel stiffness

Figure 5.5　Maximum effective shear strains in the tunnel with different tunnel stiffness

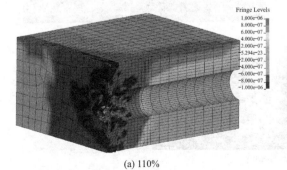

(a) 110%　　　　　　　　　　　(b) 90%

Figure 5.6　Soil liquefaction distribution in the soil with different tunnel strength (one)

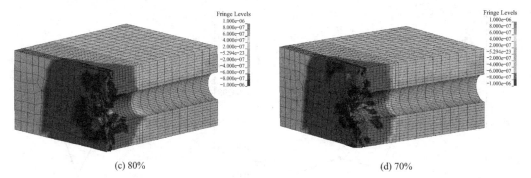

(c) 80%   (d) 70%

Figure 5.6 Soil liquefaction distribution in the soil with different tunnel strength (two)

5.2 Bulk modulus *K* of Saturated Soil

The compressibility of saturated soil under large compressive loading is determined by the compressibility of pore water and soil particles. Particularly, in most cases, air bubbles may exist in the pore water, which significantly increases the soil compressibility. The bulk modulus of soil K was therefore changed to 300MPa, 600MPa and 800MPa to check the influence of soil compressibility on tunnel response. Figure 5.7 shows the damages of the tunnel lining with different K's under the blast loading of 100kg TNT. With small bulk modulus of soil, the failure mode was governed by the tensile strength of the material instead of the phase lag of vibration. When K was larger than 600MPa, no facture occurred and only some elements in the outside surface of the tunnel failed. In the 800MPa case, the damage was even smaller. Figure 5.8 shows the relationship between K and the failure area in the tunnel while Figure 5.9 shows the relationship between K and the maximum plastic strain in the lining. The damage of tunnel was smaller with larger K. Then the stiffer soil provided more support to the tunnel lining and protected it from severe failure.

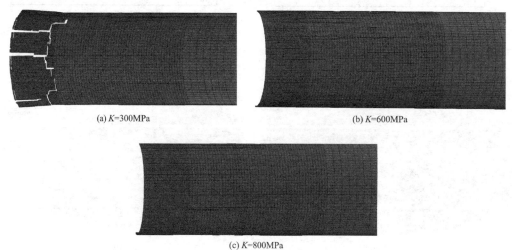

(a) K=300MPa   (b) K=600MPa

(c) K=800MPa

Figure 5.7 Failures in the tunnel of 100kg TNT with different bulk moduli of soil
(300MPa, 600MPa and 800MPa)

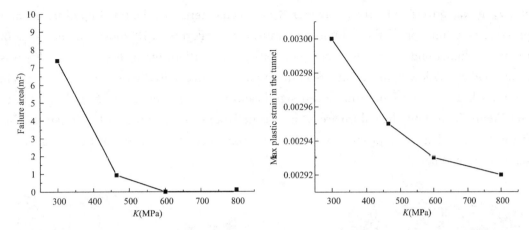

Figure 5.8 Failure areas in the tunnel of 100kg TNT with different soil strength K (300MPa, 475MPa, 600MPa and 800MPa)

Figure 5.9 Maximum plastic strains in the tunnel of 100kg TNT with different soil strength K (300MPa, 475MPa, 600MPa and 800MPa)

The volumes of liquefied soil and maximum effective shear strain in the soil were shown in the Figure 5.10 and Figure 5.11, respectively. The soil liquefaction distribution was shown in Figure 5.12. When the soil stiffness was small, the tunnel lining suffered severe fractures and less energy was transferred into the soil, hence soil liquefaction was less severe. However, when the soil stiffness was large, the tunnel had no fractures and then more blast energy was transferred into the soil. But the soil stiffness was large and the response of the soil was smaller than the base case.

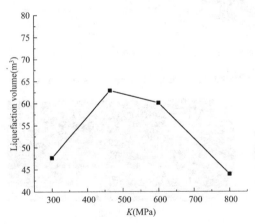
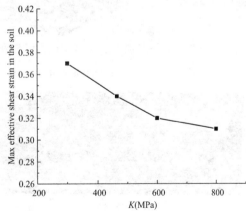

Figure 5.10 Liquefied soil volume of 100kg TNT with different soil strength K (300MPa, 475MPa, 600MPa and 800MPa)

Figure 5.11 Maximum effective shear strains in the soil of 100kg TNT with different soil strength K (300MPa, 475MPa, 600MPa and 800MPa)

5.3 Liquefaction Susceptibility of Saturated Soil under Blast Loading

Liquefaction susceptibility of saturated soil under blast loading is influenced by particle crushability of soil and compressibility of pore water. In the numerical simulations of this study, it was governed by the parameter K_{sk}, which determines the build-up of pore water pressure as a function of the volumetric strain of soil. Figure 5.13 shows the damages of the tunnel lining under the blast loading of 100kg TNT with different K_{sk}'s. When K_{sk} was small, indicating low liquefaction susceptibility of saturated soil, there were no fractures in the tunnel. But lining failure increased significantly with an increase in K_{sk}. Figure 5.14 and Figure 5.15 show the failure areas and plastic strains in the lining with different K_{sk}'s. It can be seen that soil liquefaction significantly influenced the damage of lining under the same blast loading.

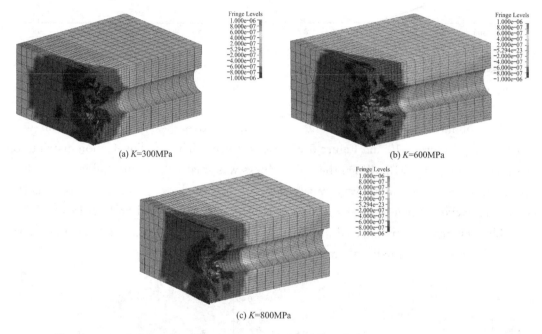

Figure 5.12 Soil liquefaction distributions of 100kg TNT with different soil strength K

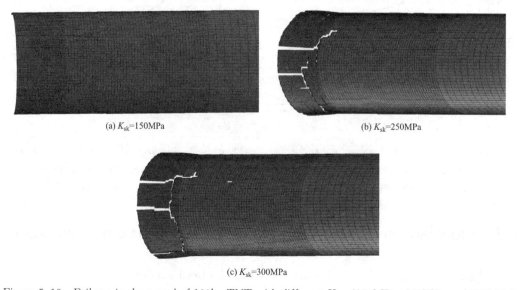

Figure 5.13 Failures in the tunnel of 100kg TNT with different K_{sk} (150MPa, 250MPa and 300MPa)

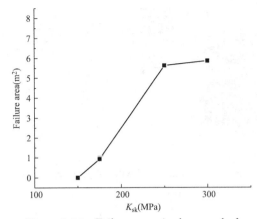
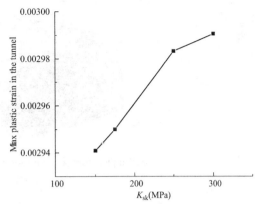

Figure 5.14 Failure areas in the tunnel of 100kg TNT with different K_{sk} (150MPa, 175MPa, 250MPa and 300MPa)

Figure 5.15 Maximum plastic strains in the tunnel of 100kg TNT with different K_{sk} (150MPa, 175MPa, 250MPa and 300MPa)

Figure 5.16 and Figure 5.17 show the liquefied volume and maximum effective shear strains in the soil with different K_{sk}'s. Larger K_{sk} generated more liquefaction and more severe deformations in the soil as shown in Figure 5.18.

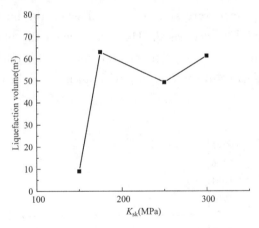
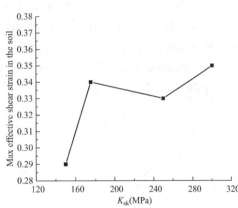

Figure 5.16 Liquefied soil volume in the soil of 100kg TNT with different K_{sk} (150MPa, 175MPa, 250MPa and 300MPa)

Figure 5.17 Maximum effective shear strain in the soil of 100kg TNT with different K_{sk} (150MPa, 175MPa, 250MPa and 300MPa)

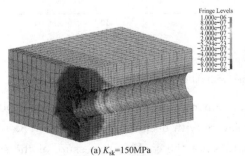
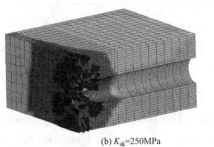

(a) K_{sk}=150MPa

(b) K_{sk}=250MPa

Figure 5.18 Soil liquefaction distribution in the soil of 100kg TNT with different K_{sk} (150MPa, 250MPa and 300MPa) (one)

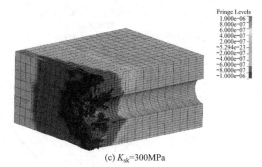

(c) K_{sk}=300MPa

Figure 5.18 Soil liquefaction distribution in the soil of 100kg TNT with different K_{sk}
(150MPa, 250MPa and 300MPa) (two)

5.4 Burial Depth

The burial depth was enlarged to 9m, 10.5m and 12.5m to investigate its influence on the tunnel response. Two blast loadings, 200kg TNT and 100kg TNT, were investigated.

Figure 5.19 to Figure 5.22 show the simulation results with different burial depths due to the blast loading of 200kg TNT. The burial depth only had slight influence on the response of the tunnel. The failures in the tunnel were almost identical when the burial depth increased, although soil liquefaction slightly decreased. However, when the blast loading was smaller, the damage of lining was much smaller with an increase of burial depth. The volume of soil liquefaction also decreased considerably with an increase in the burial depth.

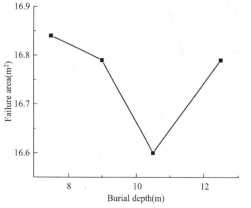 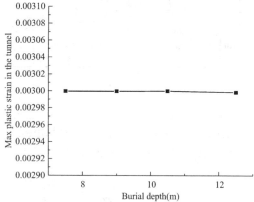

Figure 5.19 Failure areas in the tunnel with different burial depths due to 200kg TNT

Figure 5.20 Maximum plastic strains in the tunnel with different burial depths due to 200kg TNT

Figure 5.23 to Figure 5.26 show the simulation results with different burial depths due to 100kg. The fracture of tunnel was less severe with deeper burial depths. The soil liquefaction were reduced with larger burial depth since the confining pressure of the soil increased with deeper burial depth.

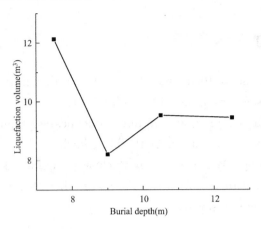

Figure 5.21 Liquefied soil volume with different burial depths due to 200kg TNT

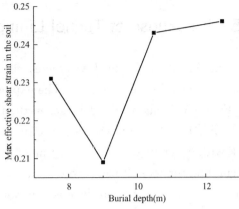

Figure 5.22 Maximum effective shear strain in the soil with different burial depths due to 200kg TNT

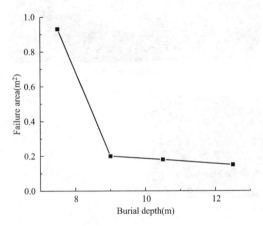

Figure 5.23 Failure areas in the tunnel with different burial depths due to 100kg TNT

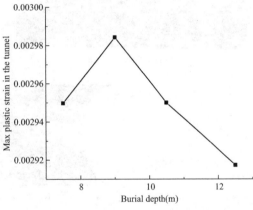

Figure 5.24 Maximum plastic strain in the tunnel with different burial depths due to 100kg TNT

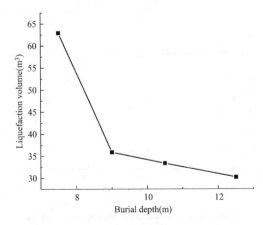

Figure 5.25 Liquefied soil volume with different burial depths due to 100kg TNT

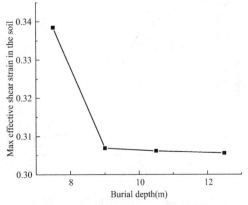

Figure 5.26 Maximum effective shear strain in the soil with different burial depths due to 100kg TNT

5.5 Thickness of Tunnel Lining

The thickness of tunnel was increased to 8cm, 10cm and 12cm while the thickness of tunnel was 6cm in the base case. The explosive was assumed to 200kg TNT.

Figure 5.27 shows the damages in the tunnel lining with different thicknesses. Figure 5.28~Figure 5.31 show the response of tunnel and soil under the blast loading of 200kg TNT with different thicknesses of tunnel lining. When the lining thickness was increased to 8cm, the tunnel still had severe damage, but the failure area was smaller than that with 6cm thickness, and the soil liquefaction was not as severe.

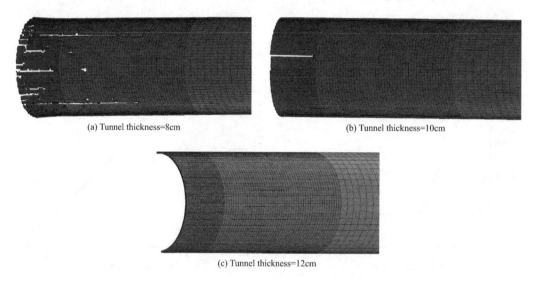

Figure 5.27 Failures in the tunnel with different thicknesses of tunnel

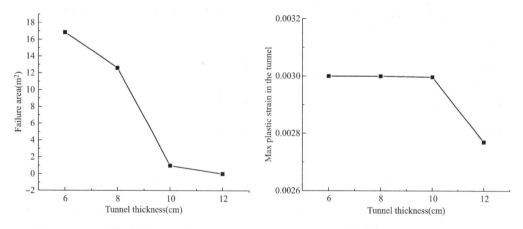

Figure 5.28 Failure areas in the tunnel with different thicknesses of tunnel

Figure 5.29 Maximum plastic strain in the tunnel with different thicknesses of tunnel

But when the lining thickness was increased to 10cm, the lining fracture was much smaller, and there was no lining fracture when the lining was as thick as 12cm. However, the lining might vibrate more significantly with an increase of the lining thickness, and might propagate more energy to the surrounding soil, unless the lining was adequately thick to resist the blast loading. With a lining thickness of 12cm, the deformation and vibration of tunnel were smaller, and much less energy was transferred into the soil. The soil liquefaction was less severe than the 10cm case.

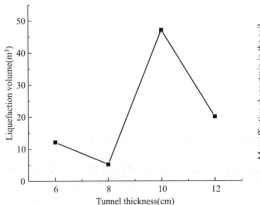

Figure 5.30 Liquefied soil volume in the soil with different thicknesses of tunnel

Figure 5.31 maximum effective shear strain in the soil with different thicknesses of tunnel

5.6 Damping of Tunnel Lining

The damping matrix in Rayleigh damping is defined as:

$$C = \alpha M + \beta K \quad (5.1)$$

Here C, M and K are the damping, mass, and stiffness matrices, respectively. The constants α and β are the mass and stiffness proportional damping constants.

In this study, tunnel lining was assumed to be continuous, while in reality it is fabricated from lining segments, which are connected by bolts and joints. The joints and the adjacent rubber that is usually employed to prevent seepage may increase the lining damping. The influence of lining damping was therefore investigated. The blast loading was assumed to be from 100kg TNT. The lining stiffness damping ratio was assumed to be 0%, 2%, 5% 7.5% while all other parameters remained the same.

There was limited fracture in the lining when the damping ratios were 0% and 2%. When the damping ratio increased, there was no fracture. Figure 5.32 shows the damage in the lining with different damping ratios due to 100kg TNT. The damage reduced slightly with an increase of lining damping. Table 5.1 shows the comparisons of some responses with different damping ratios. It can be seen that damping ratios had significant influence on the soil response. The volume of liquefied soil and the soil strain both decreased considerably with an increase in the damping ratio of tunnel lining.

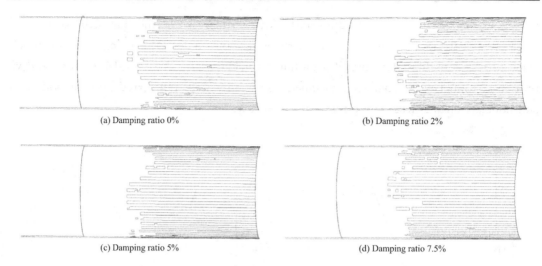

(a) Damping ratio 0%   (b) Damping ratio 2%

(c) Damping ratio 5%   (d) Damping ratio 7.5%

Figure 5.32 Failure in the lining with different damping ratios due to 100kg TNT

Table 5.1 results of simulation with different damping ratios due to 100kg TNT

Simulation results (%)	7.5	5	2	0
Failure area (m²)	0	0	0.7	1.1
Liquefaction volume (m³)	25.78	37.44	53.25	61.57
Max shear strain in the tunnel	0.00276	0.00293	0.00294	0.00296
Max effective shear strain in oil (%)	29	31	32	34

5.7 Influence of Ventilation Shaft

Ventilation system is one of the most critical considerations in transportation tunnel design. The ventilation system is helpful to control the temperature, humidity and circulation of fresh air. It can be achieved by proper sized ventilation shafts at different locations. The existence of ventilation shaft in close proximity of blast loading would modify the pattern of blast pressure propagation and change the response of the tunnel lining.

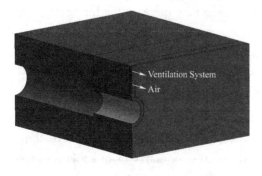

Figure 5.33 Simulation model with ventilation system in LS-DYNA

Simulations were conducted with ventilation system to investigate the influence of ventilation shaft on the responses of tunnel lining and saturated soil due to blast loading. As shown in Figure 5.33, the ventilation shaft was assumed to be of 1m×1m in size. Blast loading due to 50kg TNT, 100kg TNT and 150kg TNT was assumed to occur next to the shaft at the center of the tunnel.

Figure 5.34 ~ Figure 5.36 shows the peak pressure and impulse at different incident angles of blast loading on the lining close to the ventilation shaft. The peak pressure and specific impulse on the lining reduced significantly in the area close to the ventilation shaft, but the values were almost identical to the ones without ventilation when it was more than 1 m away. It can be seen that the ventilation shaft reduced the blast pressure only in its close proximity.

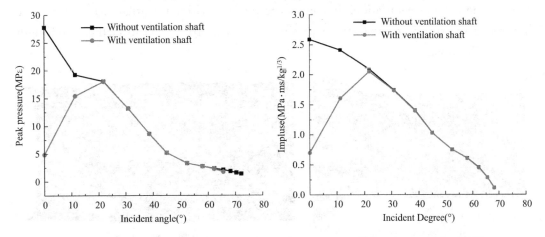

Figure 5.34 Peak pressure and impulse in the air next to lining due to 150kg TNT

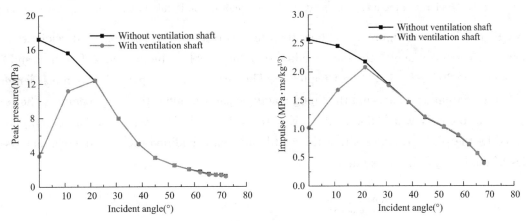

Figure 5.35 Peak pressure and impulse in the air next to lining due to 100kg TNT

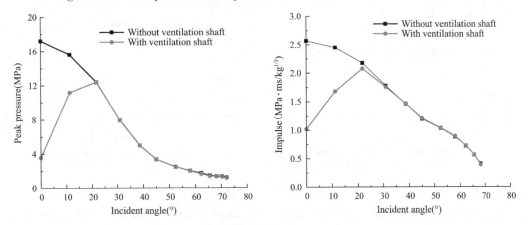

Figure 5.36 Peak pressure and impulse in the air next to lining due to 50kg TNT

The failure mode in the lining was different from that in the case without ventilation shaft. Figure 5.37 shows the comparison of failure in the lining due to the blast loading of 100kg TNT equivalent at $t=6.775$ms. Due to the existence of ventilation shaft, the lining integrity and its capacity to resist fracture was reduced. The lining had more serve fracture than the one without ventilation. In contrast to the case without ventilation, rupture firstly appeared in the section close to the explosive, and then propagated to farther distance due to lining vibration.

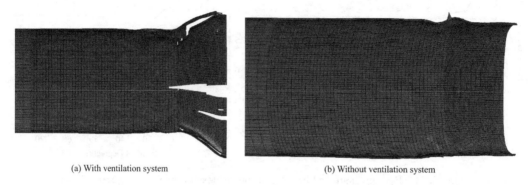

(a) With ventilation system   (b) Without ventilation system

Figure 5.37 Failures in the lining due to 100kg TNT (deformation is enlarged 20 times to view clearly)

Figure 5.38 and Figure 5.39 show the time histories of accelerations and velocities at point 1 and point 2 due to the blast loading of 100kg TNT. The locations of point 1 and 2 are shown in Figure 4.7 (a) in Chapter 4. The acceleration at point 1 was much larger with the ventilation shaft, and the lining mainly expanded under the blast loading. The responses at point 2, which was farther away from the explosive, were not considerably affected by the shaft. It appears that the ventilation shaft modified the vibration characteristics of the lining in its proximity.

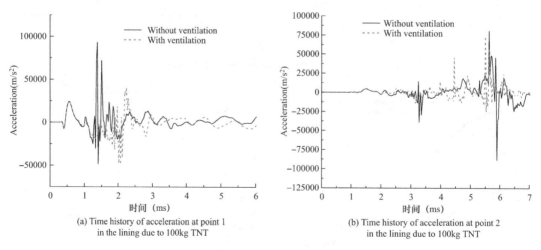

(a) Time history of acceleration at point 1 in the lining due to 100kg TNT

(b) Time history of acceleration at point 2 in the lining due to 100kg TNT

Figure 5.38 Time history of acceleration and velocity in the lining due to 100kg TNT

Figure 5.40 shows the soil liquefaction distribution at $t=6.775$ms with ventilation system due to the blast loading of 100kg TNT. The liquefied soil volume was about

12.05m³. The case without ventilation system had no liquefaction at $t=6.775$ms.

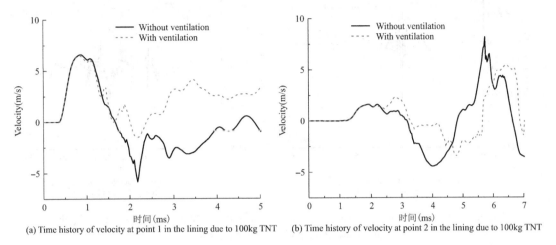

(a) Time history of velocity at point 1 in the lining due to 100kg TNT   (b) Time history of velocity at point 2 in the lining due to 100kg TNT

Figure 5.39　Time history of velocity in the lining due to 100kg TNT

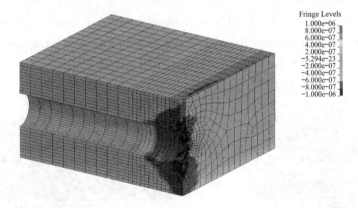

Figure 5.40　Soil liquefaction distribution at $t=6.775$ms with ventilation system due to 100kg TNT

With smaller blast loading from 50kg TNT, the lining damage with ventilation was smaller, as shown in Figure 5.41 which illustrates the failed elements on the lining surface. The bars represented the failed elements. This is consistent with the lining vibrations. The time history of acceleration at point 1 in the lining due to the blast loading of 50kg TNT is shown in Figure 5.42. Location of point 1 is shown in Figure 4.7 (a) in Chapter 4. The magnitude of acceleration reduced significantly with the ventilation system.

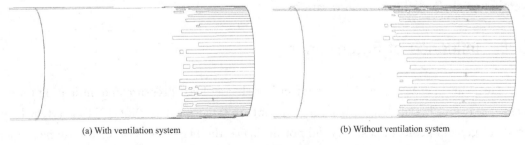

(a) With ventilation system　　　　　　　　(b) Without ventilation system

Figure 5.41　Failure in the lining with and without ventilation system due to 50kg TNT

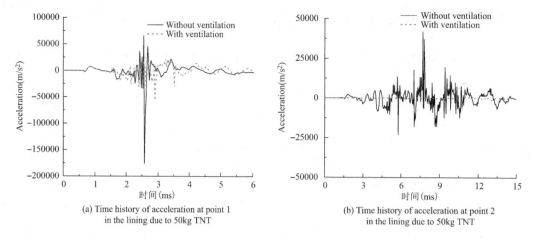

(a) Time history of acceleration at point 1 in the lining due to 50kg TNT

(b) Time history of acceleration at point 2 in the lining due to 50kg TNT

Figure 5.42　Time history of acceleration in the lining due to 50kg TNT (Location of point 1 and 2 were shown in figure 4.7a in Chapter 4)

Figure 5.43 shows the soil liquefaction distribution in the cases with and without ventilation system due to 50kg TNT. The soil liquefaction was less severe with ventilation system. The maximum effective shear strain in the soil was 28% in the ventilated case comparing to 37% in the non-ventilated case.

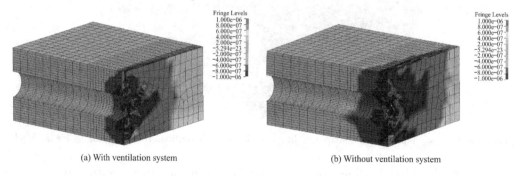

(a) With ventilation system　　　　　　　　(b) Without ventilation system

Figure 5.43　Soil liquefaction distributions with and without ventilation system due to 50kg TNT (displacement was enlarged 2 times to view clearly)

It appears that the combined effect of reduced blast pressure and modified vibration characteristic of lining was dependent on the magnitude of blast loading. With small blast loading, ventilation helped to mitigate lining damage, while with large blast loading, the ventilation shaft might even aggravate it.

5.8　Influence of Foam Liner

Geofoam liner was tested for its effect in reducing blast pressure and mitigating tunnel damage. It was found that with larger amount of explosive (⩾50kg TNT), a 4cm liner made of general geofoam basically did not influence the blast response of the tunnel. Special cases were therefore designed to investigate the attenuation of blast wave with foams due to the blast loading of 25kg TNT. The thickness of foam layer inside the lining was 4cm in prototype scale. Two densities of geofoam, 61kg/m³ and 112kg/m³, were tested. Figure 5.44 shows engineering strain-stress characteristics of foam material. Figure 5.45 shows the simulation model with embedded foam layer in LS-DYNA.

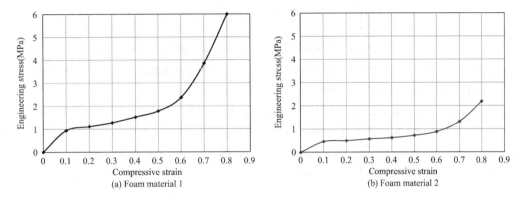

Figure 5.44 The stress-strain curve for foam material 1 and 2

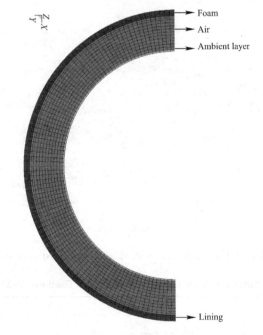

Figure 5.45 Simulation model with geofoam in LS-DYNA

The foam layer was assumed to be 10m in the longitudinal direction which was same as air domain. This was reasonable since there was no response of lining at the location farther than 10m in the longitudinal direction. The mesh size of foam layer was the same with the air domain to transfer the blast wave.

Crushable Foam constitutive model of LS-DYNA was employed to simulate the geofoam. This isotropic foam model crushes one-dimensionally with a Poisson's ratio that is essentially zero. In the implementation Young's modulus is assumed to be constant and the stress is updated assuming elastic behavior.

$$\sigma_{ij}^{\text{trial}} = \sigma_{ij}^n + E\dot{\varepsilon}_{ij}^{n+1/2}\Delta t^{n+1/2} \tag{5.2}$$

The magnitudes of the principal values, σ_i^{trial}, $i=1$, 3 are then checked to see if the yield stress, σ_y, is exceeded and if so they are scaled back to the yield surface:

$$\text{If } \sigma_y < |\sigma_i^{\text{trial}}|, \text{then } \sigma_i^{n+1} = \sigma_y \frac{\sigma_i^{\text{trial}}}{|\sigma_i^{\text{trial}}|} \tag{5.3}$$

After the principal values are scaled, the stress tensor is transformed back into the global system. The yield stress is a function of the natural logarithm of the relative volume, V, i. e., the volumetric strain.

The tunnel had no fracture in both cases due to 25kg TNT equivalent. Figure 5.46 shows damage in the lining with and without the geofoam layer. The existing of geofoam layer reduced the damage in the lining. The maximum plastic strain in the lining was 0.00288 in the case with geofoam compared to 0.00296 in the case without one as shown in Table 5.2.

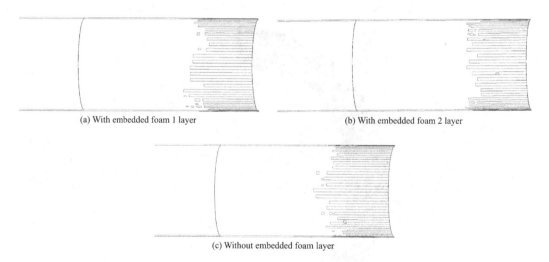

(a) With embedded foam 1 layer   (b) With embedded foam 2 layer

(c) Without embedded foam layer

Figure 5.46 Failure in the lining with and without embedded foam due to 25kg TNT

Results of simulation with and without embedded foam due to 25kg TNT   Table 5.2

Simulation results	Foam 1	Foam 2	Without foam
Length of damaged lining (m)	4	3.7	4.3
Liquefaction volume (m³)	15.75	14.89	14.23
Max shear strain in the tunnel	0.00288	0.00288	0.00296
Max effective shear strain in the soil (%)	23	23	31

The response in the soil was reduced when the foam protections were included in the model. The maximum effective shear strain in the soil was 23% comparing to 31% in the case without foam protections, although the volume of liquefied soil was basically not affected.

Figure 5.47 shows the time history of acceleration at point 1 inlining with and without foam protection. The magnitudes of acceleration in both expanding direction and shrinking direction were both reduced with the foam protection.

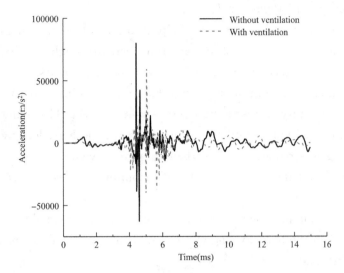

Figure 5.47 Time history of acceleration at point 1 in lining with and without foam protection due to 25kg TNT (location of point 1 is shown in Figure 4.7a)

5.9 Conclusion

A series of simulations were conducted to investigate the in fluences of different model parameters. The lining strength had obvious influences on the tunnel responses. The soil had more severe liquefaction and larger deformation with larger lining strength. Failure mode was also related to lining strength. When the strength was small, rupture firstly appeared in the domain close to the explosive, and then propagated to farther distance due to lining vibration.

Simulation results indicated that stiffer soil reduced the failure of tunnel considerably. Soil liquefaction was also less severe with stiffer soil. Liquefaction susceptibility of saturated soil had large influences on the response of tunnel and soil. The lining had more failure and larger deformation with higher susceptibility of saturated soil under blast loading.

Burial depth had very small influences on tunnel response when the amount of explosive was as large as 200kg TNT. When the explosive was assumed to be 100kg TNT equivalent, the tunnel lining had less failure and the soil had less liquefaction with larger burial depth due to larger confinement from the overburden soil.

The lining fracture decreased significantly when the lining thickness increased. When the lining thickness was assumed to be 12cm, there was no fracture due to 200kg TNT equivalent. However the soil liquefaction might be more severe with larger lining thickness. More energy was transferred into the soil when the lining had less fracture with thicker lining.

Material damping had some effects on the response of tunnel and soil. Larger damping ratio reduced the failure and deformation of the tunnel as well as the liquefaction in the soil under the blast loading of 100kg TNT.

Ventilation system with blast shafts is helpful to control the response of lining and soil due to medium blast loading. The blast wave transferred through the shafts and the pressure and impulse were reduced more than 50% near the blast shafts.

However with large amount of explosive, due to the existing of blast shafts, the rigidity of lining and the ability to resist the fracture was reduced. The lining had more serve fracture than the one without ventilation system. When the blast loading was small, the failure in the lining was reduced 43% of the one without ventilation system. The soil response was reduced and the soil liquefaction distribution was less severe with ventilation system.

Geofoam liner could provide protections to the lining due to small blast loading, but its effect disappear when the blast loading was larger than 50kg TNT for the general geofoam investigated in this study.

From the results of the parametric analysis, it can be seen that damage of tunnel lining is a result of internal of blast loading as well as dynamic interaction between tunnel lining and saturated soil. Increase of soil stiffness by certain soil reinforcement technique, such as grouting, could be adopted to reduce blast induced damage. Without significantly increasing the lining stress under static loading, overburden stress on the tunnel and its surrounding soil may also be increased to reduce the vulnerability of tunnel subjected to internal blast loading. Protective measures by geofoam which is not costly may only be effective when the amount of explosive was not large.

In this study, the joints in tunnel lining were not simulated. The joints in the lining may on the one hand reduce the strength and stiffness of the lining, but the other hand, they may increase the lining damping. From the simulation results with different lining strength, it can be seen if the strength reduction was not significant ($\leqslant 10\%$), the tunnel response was not considerably affected. However, tunnel damage was much severe when the reduction was larger. However, if lining damping could be significantly by structural measures in the joint, lining damages would be alleviated.

Chapter 6 Conclusions and Future Research

6.1 Conclusions

A series of numerical simulations were carried out in this research to study on the interaction between subway tunnels and saturated soils subjected to medium internal blast loading. This research focused on the internal explosion by medium amount of explosive (<200kg of TNT-equivalent). Numerical procedure using Finite Element program LS-DYNA that can model explosion, air-solid interaction, material damage and soil-structure interaction were used to conduct the simulations. The excess pore water pressure was studied with an existing soil model (FHWA) which can simulate pore water pressure and effective soil pressure. A new coupling method was used to study the process of blast wave propagation in the air domain inside the tunnel. Extensive parametric studies were carried out to understand the failure and damage of cast-iron subway tunnels under internal blast loading and to shed light on their protection by structural and geotechnical measures. Based on extensive numerical simulations, the following conclusions can be obtained:

- The new numerical approach combining the new blast loading scheme, centrifuge technique, air density enlargement could be used to properly simulate air blast, provided that proper element size corresponding to a specific amount of explosive is used. And as long as the free air blast at a specific scaled distance was properly simulated, the fluid-structure interaction at the same location could be properly duplicated using proper Arbitrary Lagrangian Eulerian (ALE) coupling scheme. This numerical approach can significantly reduce the number of air elements and totally eliminate the explosive elements, thus lowering the computational effort.
- The LBE method could generate the blast pressure loading accurately and transfer the blast wave to the tunnel properly. The specific impulse on the lining generated by this scheme was larger than the corresponding reflected results from UFC 3-340-02 (2008). An equivalent triangle impact load corresponding to peak blast pressure and specific impulse from Arbitrary Lagrangian Eulerian (ALE) analysis might be assumed and applied to the tunnel directly to save computer resources and the results could be acceptable.
- The failure modes of tunnel lining were different due to different amounts of explosive. With relatively large amount of explosive, severe rupture firstly appeared in the domain close to the explosive, and then propagated to farther distance due to lining vibration. The failure was governed by the tensile strength of the material.

When the blast loadings were reduced, only a little fracture occurred. The lining fracture was caused by the phase lag of vibration. This phase lag was mostly initiated by the different moments of blast loading on the lining. Overall the damage and failure of the tunnel lining was progressive in nature. The damage and failure occurred mainly during the lining vibration when the main event of blast loading was over.

- The main influence of blast loading occurred in the soil not far away from the explosive. Soil liquefaction occurred with blast loading from 50~200kg TNT equivalents. Soil had progressive failure due to both vibration of the tunnel and blast loading. With large amount of explosive and extensive lining failure, only a small portion of blast energy propagated into the soil, and the extension of soil liquefaction was actually smaller. In contrast the soil liquefactions might be more extensive when the lining failure was less severe. The lining vibrated more significantly and propagated more energy to the surrounding soil.
- Extensive parametric studies showed that damage of tunnel lining is a result of internal of blast loading as well as dynamic interaction between tunnel lining and saturated soil. Increase of soil stiffness by certain soil reinforcement technique, increasing overburden stress on the tunnel and its surrounding soil without significantly increasing static lining stress, and enhancing lining damping through structural measures at its joints, may be adopted to reduce the vulnerability of tunnel subjected to internal blast loading. Protective measures by geofoam, which is not costly, may only be effective when the amount of explosive was not large. Explosion close to ventilation shafts would reduce the blast pressure on the lining, but might not alleviate lining damage due to the reduction of lining integrity.

6.2 Future Research

Some of the future needs in this area are described in the following.
- The excess water pressure generated in the FHWA soil material model was not 100% accurate after the soil liquefied and underwent large shear deformation. Unrealistic dilation of model soil with large shear deformation led to unrealistic increase of effective stress and shear strength. More work is needed to improve the material model and the numerical scheme.
- In this study, the joints in tunnel lining were not simulated. To better simulate the effect of lining stiffness, strength and damping, it is advisable to properly simulate lining joints with proper constitutive models and finite elements.
- Lining damage due to projectiles from blast loading was not modelled in this study, which deserves further investigation.
- Foam liners could be effective in mitigating tunnel damage, but proper design of its stress-strain relationship taking into account material cost requires extensive further research.

Bibliography

[1] Blue Ribbon Panel on Bridge and Tunnel Security, Recommendations for Bridge and Tunnel Security, available online at http://www.fhwa.dot.gov/bridge/security/brptoc.cfm, 2003.

[2] UFC 3-340-02, Structures to Resist the Effects of Accidental Explosions, Dept. of the Army and Defense Special Weapons Agency, Washington DC, 2008.

[3] T C Chapman, T A Rose, P D Smith. Blast wave simulation using AUTODYN2D: a parametric study, International journal of impact engineering, 1995, 16: 777-787.

[4] X Q Zhou, V A Kuznetsov, H Hao, et al. Numerical prediction of concrete slab response to blast loading, International Journal of Impact Engineering, 2008, 35: 1186-1200.

[5] K V Subramaniam, W Nian, Y Andreopoulos. Blast response simulation of an elastic structure: Evaluation of the fluid-structure interaction effect, International Journal of Impact Engineering, 2009, 36: 965-974.

[6] J H Chung, G R Consolazio, R J Dinan, et al. Finite-element analysis of fluid-structure interaction in a blast-resistant window system, Journal of Structural Engineering, 2010, 136: 297-306.

[7] B Bjorn, B Wikman, H K Haggblad. Numerical simulations of blast loads and structural deformation from near-field explosions in air, International Journal of Impact Engineering, 2011, 38: 597-612.

[8] A De. Numerical simulation of surface explosions over dry, cohesionless soil, Computers and Geotechnics, 2012, 43: 72-79.

[9] LS-DYNA. Version 971 R6.0.0 KEYWORD User's Manual, Livermore Software Technology Corporation (LSTC), 2012.

[10] ANSYS/Autodyn-2D and 3D, Version 6.1, User Documentation, ANSYS Inc., 2007.

[11] Y Shi, Z Li, H Hao. Mesh size effect in numerical simulation of blast wave propagation and interaction with structures, Transactions of Tianjin University, 2008, 14: 396-402.

[12] L Schwer, A brief introduction to coupling load blast enhanced with multi-material ALE: The best of both worlds for air blast simulation, German LS-DYNA Forum, 2010.

[13] O C Zienkiewicz, R L Taylor, P Nithiarasu. The Finite Element Method for Fluid Dynamics, 6th edition, Butterworth-Heinemann, Oxford, UK, 2005.

[14] J von Neumann. The point source solution, in: H A Bethe, K Fuchs, J O Hirschfelder, J L Magee, R E Peierls, J von Neumann. Blast Wave, Los Almos Scientific Laboratory, Report No. LA-2000, 1958.

[15] S Marburg. Six elements per wavelength. Is that enough? Journal of Computational Acoustics, 2002, 10: 25-51.

[16] B Luccioni, D Ambrosini, R Danesi. Blast load assessment using hydrocodes, Engineering Structures, 2006, 28: 1736-1744.

[17] G F Kinney, K J Graham. Explosive Shocks in Air, Berlin and New York, Springer-Verlag, 1985.

[18] CONWEP Blast Simulation Software, US Army Corps of Engineers, Vicksburg, MS.

[19] T Børvik, A G Hanssen, M Langseth, et al. Response of structures to planar blast loads-A finite el-

ement engineering approach, Computers and Structures, 2009, 87: 507-520.

[20] S D Boyd, Acceleration of a plate subject to explosive blast loading-trial results, DSTO-TN-0270, Australia, 2000.

[21] W A Charlie, N A Dowden, E J Villano, et al. Blast-induced stress wave propagation and attenuation: Centrifuge model versus prototype tests, Geotechnical Testing Journal, 2005 28: 207-216.

[22] Liu H. Soil-structure interaction and failure of cast-iron subway tunnels subjected to medium internal blast loading, Journal of Performance of Constructed Facilities, 2012, 26: 691-701.

[23] Liu H. Dynamic analysis of subway structures under blast loading. Geotechnical and Geological Engineering, 2009, 27 (6): 699-711.

[24] Al-Qasimi E M A, Charlie W A, Woeller D J. Canadian liquefaction experiment (CANLEX): Blast-induced ground motion and pore pressure experiments. Geotechnical Testing Journal, 2005, 28 (1): 9-21.

[25] De A, Zimmie T F. Centrifuge modeling of surface blast effects on underground structures. Geotechnical Testing Journal, 2007, 30 (5): 427-431.

[26] Kutter B L, O' Leary L M, Thompson P Y. Centrifugal modeling of the effect of blast loading on tunnels. Proceedings of 2nd Symposium on the Interaction of Non-Nuclear Munitions with Structures, Panama City Beach, FL, 1985.

[27] Tabatabai H, Bloomquist D, McVay M C, Gill J J, Townsend F C. Centrifuge modeling of underground structure subject to blast loading. Technical Report, Air Forces Engineering and Services Center, Tyndall AFB, FL, 1988.

[28] Weidlinger P, Hinman E. Analysis of underground protective structures. Journal of Structural Engineering ASCE, 1988, 114 (7): 1658-1673.

[29] Kutter B L, O' Leary L M, Thompson P Y, Lather R. Gravity-scaled tests on blast-induced soil-structure interaction. Journal of Geotechnical Engineering, ASCE, 1988, 114 (4): 431-447.

[30] Davies M C R. Dynamic soil-structure interaction resulting from blast loading. Proceedings Centrifuge 94, Balkema, Rotterdam, 1994, 319-324.

[31] Helwany S M B, Chowdhury A. Laboratory impulse tests for soil-underground structure interactions. Journal of Testing and Evaluation, 2005, 33 (4): 262-273.

[32] Feldguna V R, Kochetkovb A V, Karinskia Y S, Yankelevsky D S. Blast response of a lined cavity in a porous saturated soil. International Journal of Impact Engineering, 2008, 35: 953-966.

[33] Fredrikson G, Jenssen A. Underground ammunition storages-Model tests to determine air blast propagation from accidental explosions. Technical Report, Norwegian Defense Construction Service OSLO Office of Test and Development, 1970.

[34] Davis B C. Ground shock profiles for an accidental explosion at the proposed large rocket test facilities at Arnold Engineering Development Center. Technical Report, Arnold Engineering Development Center, Air Force Systems Command, Arnold Air Force Base, TN, 1987.

[35] Coulter G A, Bulmash G, Kingery C N. Simulation techniques for the prediction of blast from underground munitions storage facilities. Technical Report. DoD Explosive Safety Board, Alexandria VA, 1988.

[36] Joachim C E, Lunderman C V. Parameter study of underground ammunition storage magazines: Results of explosion tests in small-scale models. Technical Report, U. S. Army Engineer Waterways Experiment Station, Vicksburg, MS, 1994.

[37] Hager K, Birnbaum N. Calculation of the internal blast pressures for tunnel magazine tests. Techni-

cal Report, Naval Facilities Engineering Service Center, Port Hueneme, CA, 1996.

[38] Joachim C E, McMahon G W, Lunderman C V, Garner S B. Underground ammunition storage Magazines: effects of loading density on dynamic airblast flow in small-scale models. Technical Report, U. S. Army Engineer Waterways Experiment Station, Vicksburg, MS, 1998.

[39] Chille F, Sala A, Casadei F. Containment of blast phenomena in underground electrical power plants. Advances in Engineering Software, 1998, 29: 7-12.

[40] Choi S, Wang J, Munfakh G, Dwyre E. 3D Nonlinear blast model analysis for underground structures. Proceedings of geocongress, 2006, 206.

[41] Feldgun V R, Kochetkov A V, Karinskia Y S, Yankelevsky D Z. Internal blast loading in a buried lined tunnel. International Journal of Impact Engineering, 2008, 35: 172-183.

[42] Preece D S, Weatherby J R, Blanchat T K, et al. Computer and centrifuge modeling of decoupled explosions in civilian tunnels. Technical Report, Sandia National Laboratories, Albuquerque, NM, 1998.

[43] Fragaszy R J, Voss M E. Undrained compression behavior of sand. Journal of Geotechnical Engineering ASCE, 1986, 112 (3): 334-347.

[44] Dowding C H, Hryciw R D. A laboratory study of blast densification of saturated sand. Journal of Geotechnical Engineering ASCE, 1986, 112 (2): 187-199.

[45] Bretz T E. Soil liquefaction resulting from blast-induced spherical stress waves. Technical Report, Weapons Laboratory, Kirtland AFB, NM, 1990.

[46] Bolton J M, Durnford D S, Charlie W A. One-dimensional shock and quasi-static liquefaction of silt and sand. Journal of Geotechnical Engineering ASCE, 1994, 120 (10): 1874-1889.

[47] Ettouney M M, Alampalli S, Agrawal A K. Theory of multihazards for bridge structures. Bridge Structures, 2005, 1 (3): 281-291.

[48] Koseki J, Matsuo O, Koga Y. Uplift behavior of underground structures caused by liquefaction of surrounding soil during earthquake. Soils and Foundations, 1997, 37: 97-108.

[49] Liu H, Song E. Seismic response of large underground structures in liquefiable soils subjected to horizontal and vertical earthquake excitations. Computers and Geotechnics, 2005, 32: 223-244.

[50] Yang D, Naesgaard E, Byrne P M, et al. Numerical model verification and calibration of George Massey Tunnel using centrifuge models. Canadian Geotechnical Journal, 2004, 41 (5): 921-942.

[51] Kagawa T, Sato M, Minowa C, et al. Centrifuge simulations of large-scale shaking table tests: Case studies. Journal of Geotechnical and Geoenvironmental Engineering ASCE, 2004, 130 (7): 663-672.

[52] Chou J C, Kutter B L, Travasarou T, et al. Centrifuge modeling of seismically-induced uplift for the BART Transbay Tube. Journal of Geotechnical and Geoenvironmental Engineering ASCE, 2011.

[53] Charlie W A, Dowden N A, Villano E J, et al. Blast-induced stress wave propagation and attenuation: Centrifuge model versus prototype tests. Geotechnical Testing Journal, 2005, 2892: 207-216.

[54] Abdoun, Tarek, et al. Pile response to lateral spreads: centrifuge modeling. Journal of Geotechnical and Geoenvironmental engineering, 2003, 129 (10): 869-878.

[55] Ha, Da, et al. Centrifuge modeling of earthquake effects on buried high-density polyethylene (HDPE) pipelines crossing fault zones. Journal of geotechnical and geoenvironmental engineering, 2008, 134 (10): 1501-1515.

[56] Wang Z, Lu Y, Bai C. Numerical analysis of blast-induced liquefaction of soil. Computers and Geotechnics, 2008, 35: 196-209.

[57] Nikolaevskiy V N, Kapustyanskiy S M, Thiercelin M, et al. Explosion dynamics in saturated rocks and solids. Transport in Porous Media, 2006, 65: 485-504.
[58] Jeremic B, Cheng Z, Taiebat M, et al. Numerical simulation of fully saturated porous materials. International Journal for Numerical and Analytical Methods in Geomechanics, 2008, 32: 1635-1660.
[59] Liu H. Damage of cast-iron subway tunnels under internal explosions. Proceedings Geo Frontiers 2011, March 13-16, Dallas, TX.
[60] Federal Emergency Management Agency. Reference Manual to Mitigate Potential Terrorist Attacks against Buildings. FEMA, 2003, 426.
[61] Rossum L. New York City transportation tunnels-case histories and designer's approach. Municipal Engineers Journal, 1985, 71: 1-24.
[62] Lewis B A. Manual for LS-DYNA Soil Material Model 147. Report No. FHWA-HRT-04-095, Federal Highway Administration, McLean, VA, 2004.
[63] Karagiozova D, et al. Response of flexible sandwich-type panels to blast loading. Composites Science and Technology, 2009, 69 (6): 754-763.
[64] Yi Z, et al. Blast load effects on highway bridges. I: Modeling and blast load effects. Journal of Bridge Engineering, 2013.
[65] Desai C S, Zaman M M, Lightner J G. Thin-layer element for interfaces and joints. International Journal for Numerical and Analytical Methods in Geomechanics, 1984, 8: 19-43.